# The Well-Tem

# D·I·G·I·T·A·L
# D·E·S·I·G·N

# The Well-Tempered D·I·G·I·T·A·L D·E·S·I·G·N

## ROBERT B. SEIDENSTICKER

**Addison-Wesley Publishing Company**

Reading, Massachusetts • Menlo Park, California • Don Mills, Ontario •
Wokingham, England • Amsterdam • Sydney • Singapore • Tokyo •
Mexico City • Bogotá • Santiago • San Juan

**Library of Congress Cataloging in Publication Data**

Seidensticker, Robert B., 1957–
  The well-tempered digital design.

  Bibliography: p.
  Includes index.
  1. Digital electronics.  I. Title.
TK7868.D5S42  1986       621.38′1       85–1376
ISBN  0–201–06747–1

BCDEFGHIJ-AL-89876

To Sandy and Genny

# Preface

Perhaps the best way to describe the goal of this book is to first mention a few nongoals. For instance, it is not application-specific, limited to the design of microprocessor systems, video systems, or robots. It is not an introduction to digital design, and it does not discuss case histories of previous projects. It does not teach digital design with a single logic family, and it is not a rehash of manufacturers' data books. Rather, the goal is to treat advanced digital logic design from a practical, application-independent standpoint. Organized in a casual style, similar to an experienced engineer's personal notebook, this book narrows the gap between the novice and the experienced engineer.

The practical side of digital design has received much less attention in the engineering curriculum than design theory. After all, the reasoning might go, the practical side of design is to come with experience. Naive graduates are often surprised, however, at the breadth of design in practice. Far from merely defining the chips and their interconnections, the designer must be concerned with maintainability, reliability, marketability, cost, mechanical factors, and other constraints. To name a few lower level considerations there are concerns about chip layout, noise, propagation delays, reflections, signal length, fanout, decoupling, termination, and heat. Naturally, the engineer ensures that the required function is performed by the design, but how easy is it to manufacture? Is it easily tested? How does the module interact with the rest of the system? Is the marketplace adequately understood? What techniques assist high-level design? Prototypes are too often designed without adequate concern for these constraints. The new engineer's outlook must be broadened considerably. A different approach in curriculum does not leave learning these practical aspects to chance; instead, at least, it familiarizes the student with them.

This book is a collection of "proverbs"; each proverb describes an important piece of practical digital design knowledge. The proverbs here are largely time-independent, ensuring that they remain relevant for years. They are self-contained and logically categorized for ease of reading. A thorough glossary and index make them easily referenced. The book can be read from front to back, as a textbook, or can be referenced randomly, as a handbook. References sections allow the interested reader to pursue topics further. The reader is encouraged to customize the book with annotations.

*The Well-Tempered Digital Design* is a thorough introduction to advanced, practical digital design and is intended for anyone with an interest in this area, from students to practicing engineers.

As readers embellish existing proverbs and add new ones I hope they will consider sending them to me for possible inclusion in future editions. Input from other engineers has been vital to this first edition and will continue to be so. I can be reached % Engineering Editor, Addison-Wesley Publishing Company, Reading, MA 01867.

I am indebted to the reviewers for helping to convert a rough manuscript into a polished book. Their experience and insight greatly improved this project: Jeff Rosenthal, Jeff Plotnick, Cliff Reader, Mike Moore, Linda Plotnick, and Larry Pearlstein. I also appreciate the help of my friends at Aydin Computer Systems.

RBS

# Contents

# Introduction

The disciplines of software design and digital hardware design differ in several ways. For example, software exists only conceptually (in a code listing), while hardware exists both conceptually (in logic drawings) and physically (in a circuit board). This means that not only is the reproduction cost and effort much greater for hardware, but also many physical problems such as noise, heat, device failure, and temperamental physical components, must be confronted.

On the other hand, the different software languages are analogous to hardware's different logic families. Software and hardware design use similar types of creativity. Most important to this book, however, structured design is important for both software and digital hardware. Edsgar Dijkstra's 1968 letter to *Communications of the ACM,* "GOTO Statement Considered Harmful," was one of the early stimuli pointing to the need for structured software design. Since then, there has been a flood of papers, books, and courses advocating structured software design and teaching its rules. Surprisingly, there has not been a similar phenomenon in the literature of digital design. Are there more programmers than digital designers? Is software at a more mature level than digital design? Are programmers dealing with more complex problems that need cleaner design approaches? Whatever the reasons for past neglect, this book is an attempt to partially fill the gap by addressing the topic of advanced digital design. The book emphasizes design guidelines and structured techniques.

The advice here may be somewhat contradictory; many issues have several opposing constraints acting on them. You will be encouraged, for example, to make an instruction set complete but at the same time to keep the number of instructions to a minimum. Heat sinks improve reliability but increase cost. Reducing the number of boards in a system by increasing chip density will reduce product cost but can increase testing time. The constraints are outlined here, leaving the ultimate tradeoff decisions resting with you, the designer.

# 1
## Digital Hardware
## Design Fundamentals

Students of electrical engineering, both in school and out, span a broad range of skill levels. For this book to be valuable to the largest group of engineers, it starts with a review of the fundamentals of logic design. If you are an experienced engineer, you will probably want to skip Chapter 1 entirely. Engineers who are recent graduates may need to review certain topics. Engineering students may find Chapter 1 a compact companion to a more thorough introduction to digital design. Use this first chapter as it best suits you.

## 1.1 Logic Basics

**Logic.** A distinction must be made between "logic" and "digital hardware." Logic is the theory that one learns; digital hardware is what one uses to build things. Logic has no concepts of delay or noise; digital hardware is less clean with its concern for propagation delay, crosstalk, reflections, line termination, heat, cost, and so forth. The hardware discussion here follows the presentation of digital design from a theoretical, logical standpoint.

The concept of a function is well known to everyone, even if not by this name. For example, the " + " of addition is a function that takes two numbers as input and produces a sum as output. Just as there are several basic functions with arithmetic (+, −, *, and / are the most common), there are several with logic. The first logic function is called "AND." It can be represented with symbols, like arithmetic, and it can be represented graphically. The graphical representation of a logical function is called a gate. Figure 1.1 shows an AND gate. The inputs are labelled with variable names (in this case, A and B have been chosen as the names of the input variables, and C is the output variable). A major difference between logic and arithmetic is that logical variables can have only one of two possible values. These two values can be called "on" and "off," "true" and "false," "yes" and "no," or even "red" and "blue." We will call them 1 and 0.

The truth table in Figure 1.1 defines the AND gate output given certain inputs. It shows that the AND gate outputs a 1 if both the first input *and* the second input are 1. Just as 3 + 4 is always 7, 1 AND 0 is always 0, and 1 AND 1 is always 1. Figure 1.2 shows the other basic functions. An OR gate outputs a 1 if the first input *or* the second input is a 1. A NOT gate is the only basic gate that takes just one input. It inverts the input: give it a 1 and it outputs a 0; give it a 0 and it outputs a 1. "$\overline{A}$" is the representation for NOT A. The NOT gate can be attached to the output of the AND and OR gates to produce the NOT AND (NAND) and NOT OR (NOR) gates. Note that the circles on the outputs of the NAND and NOR gates are the only graphical differences between them and the AND and OR gates.

The exclusive-OR gate (XOR) is a little different. It *excludes* the case of two 1 inputs from receiving a 1 output (the regular OR is occasionally called "in-

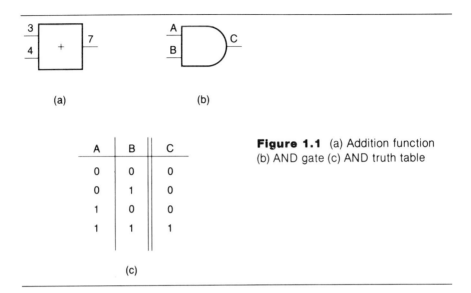

**Figure 1.1** (a) Addition function (b) AND gate (c) AND truth table

| A | B | C |
|---|---|---|
| 0 | 0 | 0 |
| 0 | 1 | 0 |
| 1 | 0 | 0 |
| 1 | 1 | 1 |

(c)

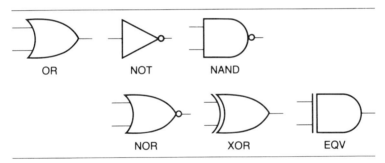

**Figure 1.2** OR, NOT, NAND, NOR, XOR, EQV gates

clusive OR" to stress this distinction). An XOR gate outputs a 1 if the first input *or* the second input is a 1, *but not both*. Another way to look at it is that the XOR gate outputs a 1 only if the inputs are different. English actually uses both kinds of OR. If a mother tells her child, "I'd like you to clean up your room or take out the trash," she uses an inclusive OR. She would surely approve of the child performing both chores and an "or both" is implied at the end of her sentence. On the other hand, if she says, "I'll buy you some ice cream or some candy," she is likely using an exclusive OR. She probably does not plan to buy both and a "but not both" is implied. The Latin language has separate words for the two or's to avoid confusion.

The equivalence function (EQV) is not commonly used. Alternatively, called exclusive-NOR, it outputs a 1 only if the two inputs are equivalent (both 0 or both 1).

The NAND and NOR gates are each sufficient to represent all the basic gates. Figure 1.3 shows a few of the possible equivalences with NAND gates. Truth tables are valuable for verifying these equivalences.

As Figure 1.3 suggests, gates can be combined to form more complex circuits (a "circuit" is an interconnected collection of gates). The output of a gate can become the input to one or several other gates. The value on the line is uniform for its entire length. However, the outputs of two or more gates must never be connected together—it makes no sense to have two sources, each trying to control the value of the variable represented by the line. (Section 1.3 describes the exceptions to this rule.)

So far, a brief introduction to logic has been presented. Two branches of mathematics that deal with logic more fully must also be mentioned. The mathematics of logic relating to digital design is known as Boolean logic, named after George Boole (1815–1864). Boolean logic is an equivalent representation of logic and is discussed in the next section. Logic is also valuable in proving the truth value of arguments through the consistency of their statements. Applied this way, it is called formal logic. Formal logic is not related to digital design and is not discussed.

**Simplification Techniques.** When studying logic from a purely theoretical standpoint, there seems little need for replacing a circuit with an equivalent implementation of fewer gates. Anticipating the implementation of the logic with hardware, however, it is important to implement the circuit re-

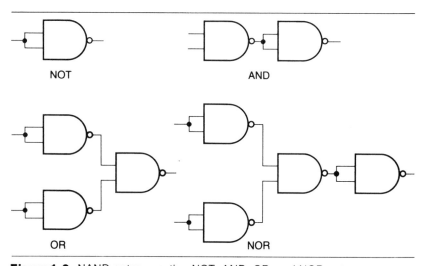

**Figure 1.3** NAND gates creating NOT, AND, OR, and NOR gates

quiring the fewest gates. As a rule, the fewer the gates, the lower the cost, the better the reliability, the less heat produced, and so forth.

The most straightforward approach is to represent the desired function with Boolean algebra and use algebraic techniques to simplify. Writing a function with Boolean algebra instead of with gates looks similar to arithmetic: a "+" is used to mean OR, a "*" means AND, and parentheses are used for grouping. Figure 1.4 shows more equivalences between logic and Boolean algebra, which should clarify the algebraic representation.

In simplifying an equation, the following theorems of Boolean algebra are helpful (they are easily verified with truth tables):

| Theorem | | Name |
|---|---|---|
| $A + \overline{A} = 1$ | $A * \overline{A} = 0$ | |
| $A + A = A$ | $A * A = A$ | |
| $A + 0 = A$ | $A * 0 + 0$ | |
| $A + 1 = 1$ | $A * 1 = A$ | |
| NOT (NOT A) = A | | |
| $A + B = B + A$ | $A * B = B * A$ | commutative |
| $A + (B + C) = (A + B) + C$ | | associative |
| $A * (B * C) = (A * B) * C$ | | associative |
| $A * (B + C) = (A * B) + (A * C)$ | | distributive |
| $A + (B * C) = (A + B) * (A + C)$ | | distributive |
| $A + (A * B) = A$ | $A * (A + B) = A$ | absorptive |
| $A + (\overline{A} * B) = A + B$ | $A * (\overline{A} + B) = A * B$ | |
| $(A + \overline{B}) * (\overline{A} + C) = (A * C) + (\overline{A} * B)$ | | |
| $(A * B) + (\overline{A} * C) = (A + C) * (\overline{A} + B)$ | | |
| $\overline{A + B} = \overline{A} * \overline{B}$ | $\overline{A * B} = \overline{A} + \overline{B}$ | de Morgan's |

A cumbersome expression such as $(A + (B * A)) + (((B * C) + C + B) * B)$, which requires seven gates, can be reduced through algebra to $A + B$, which needs only one gate. Note that many of the theorems above exist in complementary pairs, one OR-based and one AND-based.

Algebra is easy to learn, but it can be difficult to recognize when an equation is in its simplest form. A popular technique that requires little experience to use effectively is that of Karnaugh maps.

A Karnaugh map shows the output of a circuit given any input. A truth table provides the same information, but in a much less compact and less useful form. The input variables are divided into two nearly-equal sets. The possible values of one set are enumerated over the top of a grid in Gray code form, with the values of the other set along the left edge (see Figure 1.5). The Gray code changes one and only one bit as it sequences from value to value, and is shown in Appendix 1. The 0 and 1 values inside each grid cell are the desired output, given the inputs defined by the row and column in which the cell resides. If a cell could be either 0 or 1 (the application doesn't care what output is produced), put an X in it. Xs are "don't cares." The grid can be filled in with values computed from a less-than-optimal equation, and the Karnaugh map can produce the optimal equation.

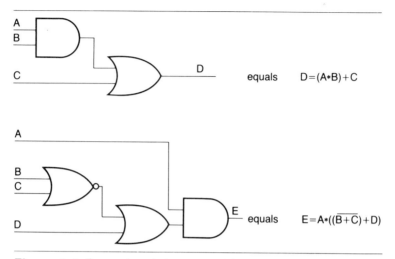

**Figure 1.4** Comparison between graphical and algebraic representation

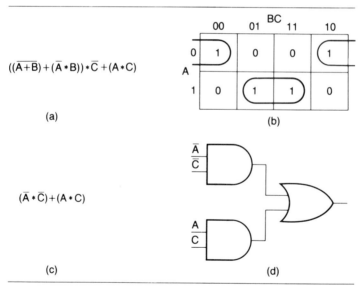

**Figure 1.5** Use of a Karnaugh map. The original equation (a) is reduced to a Karnaugh map (b) in which rectangles are grouped to reduce as many input variables to "don't cares." Each rectangle is rewritten in algebraic form (c) and ORed together to form a minimal equation equivalent to the original. This equation is converted to logic (d).

To produce this optimal equation, carefully circle rectangles of cells containing 1s or don't cares. The rectangles must each have a power of 2 as its length and width. Use rectangles that are as large as possible and yet contain only 1s and Xs. The rectangles can overlap and can wrap around the edges of the grid. After every 1 is inside a rectangle and every 0 is outside all rectangles, the simplified equation can be derived. This is the power of the Karnaugh map. Each rectangle converts to a single term in the final equation. Find the term that would create exactly those inside each rectangle (each term will be an AND combination ["product"] of several variables). Note that the bigger the rectangle, the smaller the number of variables in each term (and the simpler the final equation). These terms representing the individual rectangles are combined with ORs into the final composite equation. This is called a "sum of products" form.

**Codes and Number Systems.** The term "digital," in its broadest sense, means a noncontinuous number system. To draw a comparison between digital and continuous number systems, imagine a stairway of ten steps, each rising one foot, next to a ramp rising the same ten feet. An object could be placed on the ramp so that its lowest point was any desired distance from the floor, up to ten feet. On the other hand, the object could be placed on a step of the stairway, to raise its lowest point only a whole (integer) number of feet. The ramp is the analog to the continuous system, while the stairs are similar to the digital system.

Our familiar base ten (decimal) integers form a digital number system. This system has only ten digits but represents numbers greater than nine with different digit positions, each one being ten times more important than the one immediately to its right. The number 762, for example, is understood to represent $7 \times 10^2 + 6 \times 10^1 + 2 \times 10^0$. In general, a number in base b is understood as $a_{n-1}b^{n-1} + a_{n-2}b^{n-2} + ... + a_0b^0$, where the $a_i$'s are the digits in the number.

Logic, able to express only the digits zero and one, is a binary (two-valued) system and can use the base-two number system to represent numbers. A binary digit is called a bit. Any number of logic variables (the As, Bs, and Cs in Figure 1.2 are logical variables) can be grouped to represent a binary number. Of course, some logic variables are meant to represent a number, while others were defined for totally different purposes and make little sense grouped as a number. Suppose logic variables A, B, C, and D could be grouped as a number, with A as the most-significant bit (MSB) and D as the least-significant bit (LSB). If these variables were 1, 1, 0, and 1, respectively, the binary number would be 1101. This is converted into a decimal number as follows: $1 \times 2^3 + 1 \times 2^2 + 0 \times 2^1 + 1 \times 2^0 = 1 \times 8 + 1 \times 4 + 1 = 13$. If the variable wants to be interpreted in the opposite direction — that is, if D was the most-significant bit and A the least — the number would be 1011. This value is converted into decimal: $1 \times 8 + 1 \times 2 + 1 = 11$ (eleven).

Binary requires more digits to represent its values than number systems with higher bases, so binary is often converted into octal (base 8) or hexadecimal (base 16). Hexadecimal is also called "hex." Because 8 and 16 are powers of 2 ($8 = 2^3$ and $16 = 2^4$), binary digits can be easily grouped for the conversion to these bases. One octal digit can be converted into a 3-digit binary number, and one hex digit can be converted into a 4-digit binary number. Conversely, 3-digit and 4-digit binary values can be converted into single octal and hex digits, respectively. Since there are 16 digits in hex, the six values 10–15 need single-character representations; the six letters A–F were chosen. The following example illustrates the representation of a number in the 3 systems:

$$\begin{array}{cccccc} 1 & 2 & 7 & 2 & 2 & 6 \quad \leftarrow \text{octal number} \\ \overbrace{1\,010} & \overbrace{1\,11\,0} & \overbrace{1\,0\,01} & \overbrace{0\,1\,1\,0} & & \leftarrow \text{binary number} \\ A & E & 9 & 6 & & \leftarrow \text{hex number} \end{array}$$

This example shows that $1010111010010110_2 = 127226_8 = AE96_{16}$. Let's convert the hex value to decimal: $A \times 16^3 + E \times 16^2 + 9 \times 16^1 + 6 \times 16^0 = 10 \times 4096 + 14 \times 256 + 9 \times 16 + 6 \times 1 = 44694_{10}$. In practice, octal and hex representations are used to convert a string of binary digits into shorter and more easily managed values. Appendix 1 shows the conversion between decimal, binary, octal, and hex values.

A "byte" is a collection of bits. The original definition of byte said that the number of bits that formed the byte was variable, to be defined by the user. Now, a byte is defined almost universally, as exactly eight bits long. Another important definition is that of "K." While "k" is the metric prefix for 1000, K means $2^{10}$, which equals 1024. For example, 8 km = 8000 meters, but 8K bytes = 8192 bytes. Note that k is a prefix to the units, but K is often a suffix to the number itself.

Negative numbers are usually represented in 2's complement notation. To use this notation to represent the negative of a number, decide on the number of bits available to hold the number, invert each bit of the number (including initial zeros), and add one. For example, say there are eight bits in which to represent the value $-3$. We represent $+3$ as 0000 0011, invert each bit to get 1111 1100, and add 1 to get 1111 1101, which is the 2's complement representation of $-3$. If there are 12 available bits, we would represent $-3$ as 1111 1111 1101 (the spaces dividing these binary numbers into 4-bit pieces are only inserted for clarity). Subtraction is done by taking the two's complement of one number (negating) and adding it to the other. When performing base 2 additions, do not forget that $1 + 1 = 10$; in other words, $1 + 1 = 0$ and a carry to the next place. For example:

$$\begin{array}{ccc} 5 & 5 & 101 \\ \underline{-9} & \underline{+(-9)} & \underline{+1111\ 0111} \\ -4 & -4 & 1111\ 1100. \end{array}$$

Since the most significant bit in the binary result (1111 1100) is 1, it must be a negative value. Let's negate it (invert and add one) to see what value it has: it becomes 0000 0100 = 4. Therefore, 5 − 9 = −4. Here is another:

$$
\begin{array}{ccc}
11 & 11 & 1011 \\
-\ 6 & +(-6) & 1111\ \ 1010 \\
\hline
5 & 5 & 1\ \ 0000\ \ 0101.
\end{array}
$$

When dealing with 8-bit values, the leftmost binary digit of the result is lost, leaving 101 = 5 as the result. Having an extra bit in the result, as in this example, is called overflow.

One's complement notation represents negative numbers as the logical inverse of the corresponding positive number; that is, it is the same as two's complement notation except for adding the 1. It is less often used (there are two representations for zero), but numbers can be negated with hardware much faster. With both schemes for representing negative numbers, a negative number can be easily detected by the most significant bit being a 1.

There are other counting schemes than the base 2 number system. The Binary Coded Decimal (BCD) scheme provides a simple conversion between decimal and binary representations: each four-bit group represents exactly one *decimal* digit. The Gray code is valuable in certain applications because only one bit changes between successive values, unlike base two counting. Appendix 1 ennumerates these codes.

Several codes have been developed for the encoding of text. The most widely used is the American Standard Code for Information Interchange (ASCII). This is a seven-bit code and, as such, can represent $2^7 = 128$ characters, including upper- and lower-case letters, numbers, punctuation marks, and special "control codes." Control codes handle communications protocols and such formatting functions as carriage return and backspace. Another popular code is EBCDIC, an eight-bit code developed and used by IBM. It represents the same characters as ASCII plus extra punctuation marks and control codes.

**Multi-Gate Functions.** Several standards exist for the drawing of logic devices. The most important is ANSI Y32.14, sponsored by the American National Standards Institute. This standard governs the drawings in this book. More information can be obtained from the Institute of Electrical and Electronics Engineers (IEEE) or from some of Texas Instruments' more recent data books, including the *1981 Supplement to the TTL Data Book*. The IEEE is located in New York, NY and Texas Instruments, Inc. is located in Dallas, TX.

Gates, as seen above, can be combined to produce more complex functions. It is important to understand the most common ones, label each as implementing a certain function, and in the future, treat them as black boxes. A black box is an entity performing a function. If we know what a module's inputs and outputs are, we often want to simplify things by ignoring *how* it works and concentrating on what it does. Example implementations with gates can be found in manufacturers' logic data books.

Logic diagrams for several multi-gate functions are shown in Figures 1.6 and 4.6. A multiplexer selects one input out of many. There are eight inputs, labelled from 0 to 7. The A, B, and C inputs taken together are interpreted as a binary number from 0 to 7. The input signal selected by this value is routed to the output. A demultiplexer, given a binary number as an input, enables (assigns a 1 to) only the selected output. All unselected outputs are given the value 0. The difference is that the multiplexer outputs one signal *from* many inputs while the demultiplexer outputs *to* many signals. A magnitude comparator compares the values of two numbers and decides which is greater. An adder, as might be expected, arithmetically adds two numbers. Finally, an Arithmetic/Logic Unit (ALU) can perform any of a number of operations on two numbers, including add, subtract and logical AND.

When complex functions are designed with gates, a valuable distinction becomes important: that between positive and negative logic. De Morgan's theorem states that $\overline{A * B} = \overline{A} + \overline{B}$ (see the table in Simplification Techniques, above). This can be easily proven with truth tables and is graphically represented in Figure 1.7a.

A circuit uses positive logic when it uses only AND and OR gates (i.e., no inversions of any kind are made). Negative logic implies the use of gates with inverting circles on all inputs and outputs. A negative logic 1 (true) value is the same as a positive logic 0 value and is the opposite of a positive logic 1 value. The two ways of expressing the same function as shown by de Morgan's theorem are an example of the equivalences possible. $A * B$ uses positive logic; $\overline{A} + \overline{B}$ is in negative logic. Practical design uses mixed logic—that is, positive and negative logic are each used where it works best. Figure 1.7 shows an example.

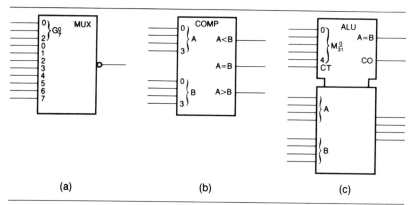

(a)  (b)  (c)

**Figure 1.6** Logic symbols for (a) multiplexer, (b) comparator, and (c) ALU

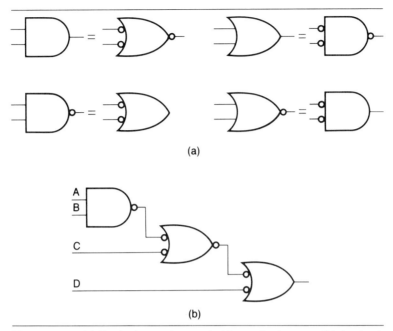

**Figure 1.7** (a) Equivalence due to de Morgan's theorem (b) A mixed logic circuit used to evaluate $(A * B) + \overline{C} + \overline{D}$

## 1.2 Advanced Logic

**Devices with State.** The multi-gate functions discussed so far are fairly simple. The outputs are functions of the inputs. The simplest device with state (memory) is the S-R (set-reset) latch, shown in Figure 1.8. The outputs, as seen from the truth table, are not merely functions of the inputs, but can depend on previous events ($Q_n$ is the current Q output; $Q_{n-1}$ is the previous output). A 0 pulse on the $\overline{S}$ input sets the latch (causes Q to become 1), and a 0 pulse on the $\overline{R}$ input resets the latch. A 0 pulse is a part of a signal that holds a 1 value, goes to 0, and returns to 1. Typical operation rarely has both the $\overline{S}$ and the $\overline{R}$ inputs active (equal to 0) simultaneously. It is simple to show that two NOR gates in an identical configuration also produce a latch, but here the set and reset pulses must be 1 pulses.

Various problems were eliminated from and improvements added to the simple latch to arrive at the edge-triggered, master-slave flip flop. Figure 1.9 shows a D (data) flip flop, its state table, and its timing diagrams. Manufacturer's data books usually provide circuit diagrams of the internals of flip flops and other devices. This flip flop still has its $\overline{S}$ and $\overline{R}$ inputs but the clock

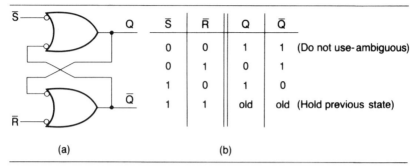

| $\bar{S}$ | $\bar{R}$ | $Q$ | $\bar{Q}$ | |
|---|---|---|---|---|
| 0 | 0 | 1 | 1 | (Do not use- ambiguous) |
| 0 | 1 | 0 | 1 | |
| 1 | 0 | 1 | 0 | |
| 1 | 1 | old | old | (Hold previous state) |

(a)　　　　　　　　　　　　(b)

**Figure 1.8** (a) S-R latch diagram and (b) Truth table

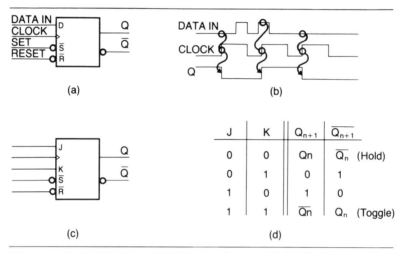

(a)　　　　　　　　　　　　(b)

| $J$ | $K$ | $Q_{n+1}$ | $\overline{Q_{n+1}}$ | |
|---|---|---|---|---|
| 0 | 0 | $Qn$ | $\bar{Q}_n$ | (Hold) |
| 0 | 1 | 0 | 1 | |
| 1 | 0 | 1 | 0 | |
| 1 | 1 | $\bar{Qn}$ | $Q_n$ | (Toggle) |

(c)　　　　　　　　　　　　(d)

**Figure 1.9** (a) D flip flop with (b) timing diagram. (c) J-K flip flop—two inputs and more generality than the D flip flop. (d) Truth table showing transitions after the next rising clock edge

and D inputs are typically more valuable. The timing diagram is a graph of the signal levels vs. time; a signal can be either low (0) or high (1) and time increases to the right. The appropriate edge (transition) of the clock (the rising edge, in this case) transfers the value of the D input to the output. By comparison, a "latch" is level triggered. Note the arrows in the timing diagram showing the who-caused-what sequences. These are optional but are often included to clarify the causes of nonobvious signal transitions. To summarize, the D flip flop is used to store (remember) a single bit of data. The D input is connected to the signal to be stored and the edges of the clock tell when the D input is to be sampled. The data at the input at the sampling instant is transferred to the output (sometimes labelled as "Q").

The J-K flip flop provides a superset of the capabilities of the D flip flop. As well as being able to clock in a high or low value, it can toggle (invert its present value) or hold (do nothing). A J-K flip flop can easily be converted to act like a D flip flop.

Now that devices with state have been introduced, here are a few more definitions. "Combinatorial" design uses only gates, multiplexers, and other devices that have no state. Synchronous design not only uses devices with state but must have them all in phase with a single master clock. That is, everything operates in step, like a marching band. Asynchronous design may use devices with state, but does not use a master clock. Any signal that has transitions can serve as a clock in an asynchronous design. Properly used, asynchronous is like a relay race. Different sections of the design perform their functions and pass the results on to other sections, much as relay runners pass along the baton. Each section runs at its own pace and communicates with its successor only when done. Any extra delay in one section (an extra long communication cable, for example) is easily handled by asynchronous design's adaptive nature.

Synchronous design is usually the preferred technique, although asynchronous communication paths often encourage asynchronous designs. Properly used, asynchronous design can work well. The main danger is in improperly designed synchronous circuits which carelessly throw in *asynchronous* techniques here and there. The two design techniques do not usually mix well.

With flip flops included with our building blocks, more complex functions are possible. A register is a group of perhaps four or eight D flip flops, each clocked with the same clock input. An entire byte of data might be stored in a single device (black box). Shift registers perform a superset of a register's function. As well as storing data, they can shift it. For example, if 1101 was loaded into a four-bit shift register and was shifted right once, the rightmost 1 would be lost, the other three bits would slide to the right, and a 0 would become the new leftmost bit: 0110.

Counters are quite valuable. With each cycle of the clock, they increment and output the value they contain. Some can also decrement. A four-bit counter might count from 0 to 15, or it might be a BCD (decimal) counter and count from 0 to 9. Note that some bit combinations (from 10 to 15) are valid BCD values. The counter skips over these as it increments from 9 back to 0 (see Appendix 1 for the BCD codes).

The components mentioned above represent the vast majority of simple devices used in practice.

**Memory.**  The storage of data is so important that it is valuable to deal with logical devices containing large amounts of memory, much more than that found in a single register. Memories have finite size; we need to consider their depth (number of units of memory) and their width (the number of bits in each unit). For example, a 1024 x 4 memory holds 1024 units of memory and each time we request a unit of memory (a "word"), we get 4 bits. Requesting a word is called "addressing the memory." Just as the 8-bit multiplexer needs a 3-bit address to select the output ($2^3 = 8$), so a 1024-word memory needs a 10-bit address ($2^{10} = 1024$). In other words, a 10-bit binary number can have 1024 possible states and can be used as a simple, compact way of expressing the location of the desired word. Presenting the memory with an address causes the memory to retrieve the contents of the addressed locations. A memory of this sort is called a random-access memory (RAM) because any memory word can be accessed as easily as any other word. Imagine the memory like many labelled shoeboxes, each holding a number as its contents. Some RAMs can be both read from and written to. That is, the shoeboxes can be consulted (without modification) or their contents can be changed. Other RAMs are read-only memories (ROMs). The contents in their shoeboxes are unchangeable but can be consulted.

Memories have other characteristics. Some memories are volatile (they forget what was written to them when their power is turned off) and some are nonvolatile. Some memories are not RAMs at all — the address of the word *does* have an affect on the ease by which the data is retrieved, just as with a magnetic tape. These are called serial memories.

Devices which are partly mechanical are used for memory also. These devices can store their data on tape (not unlike an audio tape), magnetic disk (similar to a phonograph record), and video disc.

**Computers.** Digital computers are important tools being used in increasing numbers. Computers use digital logic to execute instructions (software). Each machine has an "instruction set" — the set of instructions that it can perform. Typical instructions are "add these two values and put the result there," "perform the logical AND of these two values," "skip over the next instruction," "increment this value," and so forth. The instruction set designed anticipates the probable application(s). A high-speed scientific computer needs multiply and divide instructions and might also want to execute operations on groups of numbers (vectors or arrays). A computer used for control applications might not need the powerful arithmetic instructions but might instead require a full complement of shift and logic instructions (for example, AND, OR, and XOR).

A software program is a list of instructions, selected from the instruction set and stored in memory. Software is much more easily modified than hardware. A single wire change or chip replacement with hardware might take longer than the complete reloading of a memory containing software. In systems containing both hardware and software, a tradeoff must be made between the performance advantages of hardware and the cost and flexibility advantages of software.

Most computers are classed as von Neumann machines and, as such, are all logically equivalent to a Turing machine (a Turing machine is a hypothetical machine with a certain trivially simple instruction set). This means that the von Neumann machine with a simple instruction set can perform the same tasks as a machine with the most powerful. The computer with the powerful instruction set is easier to program and might run faster but is more difficult and expensive to build. A von Neumann architecture has a single processor and control passes in a sequentially from instruction to instruction.

The instructions executed by the computer are in a form called machine language. This is the set of instructions written for execution by the machine. The instructions are composed of 1s and 0s representing commands, memory addresses, and constant values and are used just like any other units of data to address multiplexers, control an ALU, enable a signal, address memory, or perhaps be used as memory data. Software, then, is a subset of digital hardware. It specifies what parts of the hardware should act and when.

Machine language is difficult to work with, but programming can be done at a higher level called assembly language. Assembly language provides a convenient way to alphabetically define and refer to instructions. An assembler is a program which converts the assembly language text (source code) into the corresponding machine language (object code).

High level languages are even farther removed from machine language. In fact, while assembly language is strongly tied to the instruction set of its computer, programs written in high level languages are often transportable to many other computers with different instruction sets. Popular high level languages include FORTRAN (for scientific use), Pascal and PL/I (general), BASIC (hobby and educational), LISP and Prolog (artificial intelligence), and COBOL (business). All languages have their weaknesses; this explains why different communities of computer programmers have developed hundreds of languages to best suit their needs. High level languages are converted to machine languages by programs called compilers or interpreters.

Another valuable software tool is the operating system (OS). In those computer systems developed enough to have one, the operating system is the program which is running when a user's program has finished. The user tells the OS to assemble or compile a program, run a program, edit a program, or perform other tasks. Because the operating system is in charge of managing resources such as memory and peripherals, it often provides programs with access to these resources. The operating systems for some large computers are among the largest and most challenging programs ever written.

Let's return to the logic design of a computer. If a computer has an external event about which it is concerned, it can deal with it in two ways: polling or interrupts. Suppose the event is "the user typed another key on the terminal." The processor can poll the keyboard, continuously determining whether or not a new key has been typed, or it can be executing other unrelated instructions and be interrupted by the event of a key being typed. The advantage of the interrupt is that the computer can be executing meaningful instructions while it is waiting for the event, perform whatever processing is required after the event occurs, and resume its original task, all with little effort on the part of the programmer.

One common way of classifying computer architectures looks at the number of instruction and data streams. The common von Neumann architecture has a single instruction stream and operates on a single data stream (one unit of data is operated on at a time). This architecture is called a Single Instruction stream, Single Data stream (SISD) architecture. When more parallelism is called for, the Single Instruction stream, Multiple Data stream (SIMD) architecture uses one set of instructions to operate on multiple data sets with multiple processors. The same instruction is simultaneously performed by each processor. With a Multiple Instruction stream, Multiple Data stream (MIMD) architecture, the processors are executing their own unique instructions on different sets of data.

## 1.3 The Real World

**Digital Hardware.** Now that logic has been discussed from a theoretical viewpoint, let's look at the implementation of logic in hardware. Digital hardware is a subset of analog hardware, since it uses electronic components (transistors, diodes, and resistors) as its building blocks. Digital is noncontinuous (discrete), while analog is continuous (see Codes and Number Systems, above). The 1 and 0 logic values are represented as "high" and "low." The signal on a wire is high or low depending on whether its voltage is within a given high range or a low range.

Logic works with nonphysical components. The constraints imposed by the use of actual hardware are important. One important difference between hardware and logic is that hardware requires a certain amount of time to produce an output after it is given its inputs. This time, the "propagation delay," is typically a small number of nanoseconds (billionths of a second). It is not a fixed value but will fall somewhere between the device manufacturer's guaranteed minimum and maximum propagation delay times. Wire also has a propagation delay, which is usually between one and two nanoseconds per foot. Another constraint is that the data inputs to synchronous devices like registers and flip flops must be stable a certain amount of time before the clock edge arrives ( the setup time) and after the clock edge arrives (hold time). A digital device (often called a "chip") can output to only a certain number of inputs, ranging from one to dozens (this number is the "fanout").

There are many different digital hardware families. The most popular families are Transistor-Transistor Logic (TTL), Emitter-Coupled Logic (ECL), and Complementary Metal Oxide Semiconductor (CMOS). Each family makes different tradeoffs in power consumption, heat dissipation, speed, noise immunity, cost, and so forth, and is suited to different applications (discussed in Section 3.1). It is possible to use more than one family in a design by using translators to change the characteristics of the signals communicating between the different devices.

The complexity of devices falls into several different categories, as shown in Table 1.1.

Digital, since it is really analog electronic hardware, must contend with many problems that logic does not address. For example, wires carrrying a signal can emit electric and magnetic fields. If these fields alter the signal in a nearby wire, we have a problem with crosstalk. These fields on a large scale are called electromagnetic interference or radio-frequency interference (EMI/RFI). It is important to ensure that (1) external EMI/RFI does not interfere with the operation of a board or system and (2) the board or system does not generate too much EMI/RFI itself.

Wires over a certain length act as transmission lines. These have special problems not found in shorter wires. Any wire whose length is measured in feet almost certainly needs to be examined with transmission line theory in mind, but it can apply to shorter lines when high-speed logic families are used.

Imagine a 10-foot rope fastened at one end: if you send a wave down the taut rope by shaking the free end, the wave reflects off the fixed end and, with little attenuation, returns. On the other hand, if the other end were not fixed but loose, it would absorb the wave and reflect nothing. Wave reflection is a problem with transmission lines. If there is a mismatch between the impedance of the wire and the impedance of the device input at the end of the wire, there is a reflection, just as there is with the rope with the fixed end. Reflections must die out (fade) before the voltage that started at the output of the source is stable at the input of the destination. These reflections can be eliminated by mimicking the loose end of the rope: if the wire end is terminated with a resistor that shows no impedance discontinuity to the signal wavefront, it is absorbed and no reflections occur.

Other differences are apparent in the move from logic to hardware. In logic, any logical device can be imagined, but only certain devices are important enough (have enough marketability) to be manufactured as physical hardware.

| Category Name | # of Equivalent Gates | Examples |
| --- | --- | --- |
| Small-Scale Integration (SSI) | < 12 | gates, S–R flip flops |
| Medium-Scale Integration (MSI) | 12–100 | registers, shift registers, counters, multiplexers, demultiplexers |
| Large-Scale Integration (LSI) | 100–1000 | small and medium-sized memories |
| Very Large-Scale Integration (VLSI) | 1000 + | microprocessors, large memories |

**Table 1.1** Comparison of different categories of device complexity

A 16-bit counter may not exist as a single available component, but it might be possible to concatenate four 4-bit counters to create one. Devices that are often combined to make larger ones, as in this example, are usually designed with any required concatenation signals. Two 8-bit registers easily make a single 16-bit register with no communication between them. However, two shift registers need to shift bits between each other and adders need to pass along a carry bit to their more significant partners.

At the highest level of design, pure logical devices are the tools of the designer. As the design becomes closer to physical realization and the project's "Control Section" and "System Interface" are converted from simple black boxes into multiple boxes, attention is shifted to the actual hardware used in the implementation. Top-down design, of which this logical-first-physical-later philosophy is a part, is discussed in detail in Chapter 2.

A signal that contains considerable noise can be a difficult problem. A single, low-to-high transition might be construed as three edges — rising, falling, and then rising again — if there is sufficient jitter (noisiness) in the signal. The problem is that a small amount of jitter can cause the signal to cross the low/high threshold several times when only one transition was intended. The solution is to have two thresholds, a situation known as hysteresis. As a jittery rising edge crosses the low-to-high threshold (the higher threshold), the transition is made. If jitter then drops the voltage level below this threshold, no transition is made; the high-to-low threshold is a safe distance below. Hysteresis is used in a device called a Schmitt trigger and in the inputs to devices that might receive noisy signals.

The Schmitt trigger makes the analog world easier to live with. Other devices that deal with the analog world are used in digital design; let's define a few. To convert a digital value, represented by a number of bits, into an analog voltage, a digital-to-analog converter (DAC) is used. An analog-to-digital converter performs the opposite transformation. An accurate square wave for a clock is essential to synchronous designs; a crystal is the heart of the clock generating circuitry. Monostables ("one shots"), once triggered, use the charging of a user-supplied capacitor to create a pulse of a certain length. Changing the capacitance of the capacitor changes the duration of the pulse by changing the time required for the charging capacitor's voltage to reach a preselected level.

Let's return to the discussion of memories and discuss actual memory device types. Random-access memories (RAMs), as discussed above, can be both written to and read from, and read-only memories (ROMs) can be only read from. There are two kinds of volatile RAMs (RAMs that forget when the power is turned off): static and dynamic RAMs. Static RAMs store their bits in a way similar to how an S-R flip flop does — with gates. Dynamic RAMs use tiny capacitors to store their bits. The capacitors slowly lose their charge and must be periodically recharged ("refreshed"). Refreshing dynamic RAM chips does not require much time but can be inconvenient and can demand special hardware. Static RAMs can be faster and do not require refreshing, but dynamic RAMs can be made with more storage per device and usually cost less.

Read-Only Memories (ROMs) are nonvolatile and come in many forms. The contents of ordinary ROMs are defined by the manufacturer. Changing the contents of future ROMs is possible but expensive. Programmable ROMs (PROMs) are easy to use because a user can do the programming. A PROM programmer is a small desk-top device that loads ("burns") the desired contents into the PROM. Some types of PROMs can be programmed more than once. An ultra-violet PROM (UVPROM) can be erased by exposing the tiny silicon device to ultra-violet light through a small transparent window. Electrically-erasable PROMs (EEPROMs) can be erased electrically. Either PROM, once erased, can be reprogrammed. Nonvolatile RAM (NOVRAM) is a "read-mostly memory," which means that it can be written to, but only a limited number of times. Core memory is an older type of memory device that uses tiny ferrite rings (cores) to store a magnetic field. Not only is core memory nonvolatile, but it can be read to and written from randomly. It is rarely used now because of its high cost and low storage density.

Bubble memory is a nonvolatile, serial memory. Charge-coupled devices (CCDs) are also a type of serial memory but are volatile. Serial memory can be both read and written but can not be accessed randomly. Their long latency (the time to get the data stream started) and relatively slow data rate (the time per data word, once the data stream is started) makes these memories less popular than RAMs.

Several other types of memory are used in special applications. A "first in, first out" memory (FIFO), also called a "queue," is used when the characteristics of a data stream are to change. The first datum (piece of data) into the FIFO is the first one out, just as a queue (line) of people waiting to get into a theater is filled at one end and emptied at the other. No link exists between the rate of filling and that of emptying. Similarly, a FIFO can be filled by one process and emptied by another. It has limited storage, but is valuable for loosely linking two communicating processes (hardware modules). For example, a FIFO might smooth the flow between one process that supplies data in bursts and another process that consumes the data at a steady rate.

A "last in, first out" memory (LIFO), also called a "stack," can be modelled by spring-loaded stack of plates found in a cafeteria. Plates (or data) are removed from and added to the same end (the top of the stack), while both ends of the FIFO (queue) are active. Occasionally, a computer is interrupted in the middle of a task to perform a more urgent task. The computer often stores a small amount of information — enough so that it can resume where it left off — on a stack before it executes the more urgent task. This information is the *state* of the computer. The computer might get interrupted again, after which the computer would again store its current state on the stack and go execute *that* task. As each task is finished, the computer retrieves state from the stack and resumes execution on the previous suspended tasks. The stack maintains proper order of the state. FIFOs and stacks can be actual memory devices or can be implemented using random-access memories.

Logic theory has no concern for size, cost, and power dissipation, but hardware design does. Microcomputers are small and inexpensive computers built with a small number of hardware devices. Applications for microcomputers are increasing rapidly and they are extremely valuable additions to the designer's set of tools. Microprocessors are contained in microcomputers. Microprocessors perform only computations; microcomputers contain RAM, ROM, interfaces to the rest of the system, and the other necessary parts to make a complete computer, plus the computing power of a microprocessor.

Microprocessors execute instructions (software) stored in RAM or ROM. Some instructions perform arithmetic or logical processing, while others control the sequencing of instructions. If higher performance than that which can be provided by a microprocessor is needed, bit-slice processors can be used. These are different from microprocessors in two ways: first, they do *not* provide the instruction sequencing capability and second, they are more flexible design tools. Other devices are added to provide sequencing. The flexibility comes from their being bit slices: bit-slice processors can be cascaded (combined) to form computers that operate on data of any width while microprocessors have a fixed data width. For example, an 8-bit-wide microprocessor would slow down when processing 16-bit-wide data, but four 4-bit slices can be cascaded to handle the job. This cascading is similar to that in which four 4-bit counters are cascaded to make a single 16-bit counter. A separate sequencer controls the instruction stream to the bit-slice processor. While more difficult to design and usually much more expensive, the bit-slice approach is often used instead of microprocessors in high performance applications. In fact, some minicomputers are built using bit-slice processors.

Most hardware devices are standard products of the manufacturer. Some, however, are customized by the user or by the manufacturer at the user's request. The simplest type of customized device is programmable logic, which is valuable for performing certain kinds of complex combinatorial functions. The programmable logic device can be programmed to reflect the desired combinatorial equation (in a way similar to how a PROM is programmed). Programmable logic is inexpensive and can be customized quickly and easily. At the opposite extreme in time, effort, and cost, is custom logic. Here, a totally custom integrated circuit is designed to the user's specifications. A less flexible but less costly approach is semi-custom logic. The user's device can be built with predefined building blocks such as flip flops or counters. Programmable logic, semi-custom logic, and custom logic are ways of reducing cost, device count (the number of devices on a circuit board), and board size. These types of logic are defined in more detail in Chapter 2.

**Digital vs. Analog.** With all the analog considerations such as noise, propagation delay, and reflections, the digital designer is constantly reminded that digital hardware is a subset of analog hardware. However, within the abstract, comforting world of digital theory, the designer is insulated from the complexities of analog. For example, the noise and the signal in most analog systems appear the same. An analog telephone signal being transmitted over a long distance needs to be periodically amplified. Between each amplifier, however, noise accumulates, becomes part of the signal, and is amplified along with the intended signal. A digital signal also accumulates noise, but the logical signal is easily interpreted from the slightly noisy waveform that arrives at the receiver. Only the signal is passed on.

Digital provides an excellent arena for computing applications. The digital medium was strange territory when computers were first built, but the evolution of an art into computer science over recent decades has provided an extremely valuable tool. Analog computers have their uses (they solve differential equations very well), but most present computer applications had to wait for the digital computer.

**Synchronous Design.** Several techniques are used in large-scale synchronous design. A "bus" is a group of signals with more than one source. True, it does not make sense to have more than one source simultaneously determining the value of a signal, but a bus is designed so that there can be multiple sources with only one source at a time outputting data. While one source is active, the other sources are disabled. Each source outputs data when and only when it is enabled.

Busses are like multiplexers: with either, you can select one signal out of many. There are differences, however. Once several sources are bussed together, each of the potentially many destinations receives the same information — that of the solitary enabled source. Also, the other sources become inaccessible while disabled. These can be limitations and a multiplexer eliminates these situations; however, multiplexers are extra logic, while bussing does not require any logic overhead.

Two basic bussing techniques are used: wired logic and tri-state ("3-state") outputs. Wired logic comes in two forms. TTL's open collector outputs, available on certain devices, can be tied together. Any of these outputs connected together can pull the signal low, hence the name "wire-AND" for such a combination. ECL's outputs are always open-emitter, so that wire-ORing is always possible. As the name suggests, tri-state outputs have a third output state in addition to low and high, called "high impedance." The output of such a device, when in its high impedance state, does not significantly control the current on the wire, as a normal signal would. This allows another source on the same line to control the state of the line.

A synchronous design has a sequence of clocked devices (registers or flip flops) with combinatorial logic in between. The propagation delay of the devices and the clock period (the time during which the propagation is done) must be considered. Careful synchronous design is somewhat like musical chairs: after the music begins (after a clock edge comes), data propagates through the clocked devices and into and through the combinatorial devices. However, when that music stops (when the next edge comes), the data must be stable at the inputs to the next synchronous device.

In a pipelined system, a simple sequence of registers or other synchronous devices should be evident. Between the registers, combinatorial logic performs operations on the data as it flows through the pipeline (see Figure 1.10). Data propagates from stage to stage and is modified as it travels. A potentially large digital structure such as a pipeline highlights the parallelism of digital: in digital, everything works all the time, and a pipelined system can be a good way to exploit this parallelism.

With combinatorial logic performing a function (multiplication, for example), the propagation delay through the logic alone gives the speed, but with a pipelined system, the frequency and the pipeline delay are both necessary. The pipeline delay is the time from when a datum (piece of data) enters the top of the pipeline until it leaves the bottom, and the frequency is the rate at which data falls out the bottom. A comparison between a logical pipeline and a

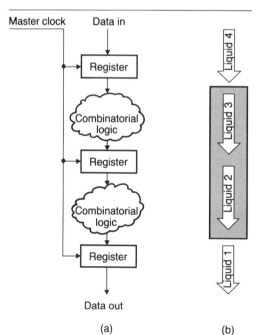

**Figure 1.10** (a) Pipelined system: data enters the top, is transformed as it moves from stage to stage, and exits at the bottom. (b) The physical analogy in which quantities of different liquids are transported by the same pipeline.

physical pipeline carrying liquid (say) is helpful: the length of the pipeline is totally independent of the rate at which liquid pours out the end, but both numbers are important.

Pipelining is used in applications where many pieces of data want a similar sequence of processing. Also, each piece of data is processed independently — that is, each does not rely on the results computed from preceding data. Not all applications can tolerate the restrictions of pipelining, but those that can are often greatly sped up with a pipelined implementation.

Many digital designs can be divided into two parts: the first part operates on data, while the second part controls the first. Controlling a complex design can be easier with the technique of microprogramming, which is the writing of microcode. A very low-level form of software (at a level lower than machine language), microcode can enable outputs onto buses, provides gate enables, and can provide data and memory addresses. Microcode is stored in RAM or ROM and must be addressed with a controller, as must any software. Of course, the use of microcode implies the need for a computer. Bit-slice processors are *always* programmed with microcode, but microcode may or may not use a bit-slice processor. Microcode is different from low-level software such as assembly/machine language in that fields (groups of data bits) in each instruction can be allocated for the control of digital signals. Assembly languages on many computers are implemented with hardware. Software such as microcode is valuable because it is easier to change and upgrade than hardware.

Different forms of memory have different speeds and costs. Computer systems typically want the fastest memory access possible. A memory hierarchy of small, fast, and expensive memory down to large, slow, and inexpensive is often an excellent way to provide reasonable speed at acceptable cost. Single registers or a small memory can form the fastest, most expensive memory. A cache memory, composed of a medium-sized amount of fast memory can be the interface to the large, slower main memory. The cache memory attempts to satisfy memory requests by mirroring recently used main memory locations. Those values found in the cache avoid a main memory access and can be provided typically five or more times faster than if the main memory access had been made.

The main memory can be megabytes in size and can be fairly slow. Because errors occasionally arise in stored data, error correcting codes are often included in this memory (adding perhaps 20–50% to the word width). As a word is read, special hardware uses the extra data in the word to ensure that it has not changed since it was written. Slower and less expensive still are magnetic media (tape or disk).

**Debugging Tools.** After the design moves from the logic sheets into a prototype, the designer can call on additional tools. A quick overview of these tools:

- Before the design is built, it can be simulated and tested with software tools. This is especially common with VLSI designs.

- The multimeter is a simple tool for testing voltages, as well as currents and resistances.

- A continuity tester lights up or sounds a buzzer to indicate a continuous electrical path (good for determining if two pins are connected).

- "DIP clips" can clip on to an integrated circuit package to provide convenient access to the signals on the leads (a DIP is a dual in-line package, the most common type of integrated circuit package).

- An oscilloscope is an expensive piece of equipment and shows the time-varying waveforms of signals.

- A logic probe can indicate the logical state of a signal.

- A logic analyzer is very expensive, but can input many signals and display their logical waveforms over time (like a timing diagram). This valuable tool is almost essential for the more complex designs.

- When the prototype is part of a computer system, the computer is often called upon to exercise the prototype.

- An excellent language for development is Forth.

**Board Manufacturing Techniques.**   There are several techniques for manufacturing circuit boards. They all use a stiff baseboard for support but use different techniques for creating the wires that carry the signals. A printed circuit (PC) board is the least expensive. Copper signal paths (traces) are laid onto the board and carry signals or provide power and ground connections. When the number of signal paths becomes more than the two sides of a PC board can handle, more layers can be added, producing a multilayer board. Multiwire is a technique in which thin wires are laid into an epoxy base. It is especially valuable for highly packed designs in small to medium quantities. Wire wrap is an often-used technique for producing prototype boards and production-volume boards in small quantity. With it, wires can be quickly and reliably changed. Boards in a system are supported by a chassis and can communicate with each other through a backplane. Backplanes can contain PC traces or wire-wrapped signals to provide the communication.

With PC, multilayer, and Multiwire boards, chip leads are inserted through holes in the board and soldered from the underside. Equipment exists to automate both the chip insertion and the board soldering. Sufficient manufacturing volume makes it worthwhile to also provide automatic testing.

While human interaction can be lessened, the personal touch cannot be eliminated totally. A silkscreen (paint on the board labelling chip locations, pin numbers, board type, and so forth) can give valuable assistance to anyone working with a board. Testability (ease of testing) is an important quality which can be given to a board only with careful thought during design.

**Integrated Circuit Manufacturers and Literature.** Manufacturers of integrated circuits provide literature and data books describing their products. A discussion of a part typically provides, among other data, the device pinout (definition of each pin), a textual description of the part, power required, heat dissipated, electrical requirements of input signals, and propagation delays of output signals. Delays are often given as typical and maximum. It is often best to ignore the typical figure: it usually does not matter what a signal usually does but it is important what limits it is guaranteed not to exceed. A logic diagram of the circuit is usually provided, but this diagram is included only to help define the function of the device and does not necessarily reflect how it is actually built. Other chapters in a data book often provide application notes and package size and shape measurements.

An important new standard for logic drawing is ANSI Y32.14. Logic symbols following this standard are often somewhat dense with information. A designer can usually obtain most of what is needed from the symbol. Usually, there is no need to study the information in the data book (see Figures 1.6 and 6.4).

# 2
## Design Considerations

At first sight, the idea of any rules or principles being superimposed on the creative mind seems more likely to hinder than to help, but this is quite untrue in practice. Disciplined thinking focuses inspiration rather than blinkers it. [1]

## 2.1 High-Level Philosophy

This section discusses the highest level of design. As a design progresses from concept through manufacture, the rules followed at each stage become more rigid. During high-level design, then, the "rules" are the most flexible and least constraining. Additionally, the designer often must make choices between somewhat conflicting demands; e.g., generality and completeness vs. sparse design and orthogonality. These conflicting demands are also discussed.

**Use Top-Down Design.** Top-down design is a rational approach to digital design. Other approaches hurry the design to the debug phase more quickly. By not sacrificing the design phase, however, top-down is the approach for completing the entire design fastest and producing the most satisfactory product. *Completely* finishing each design stage is analogous to racing a car and leaving it in each gear as long as possible. Just as the fastest car acceleration occurs with this method, so is the design is completed fastest. The steps to top-down design are:

1. **Understand the Problem.** Though this seems obvious, designs occasionally are completely debugged before a fundamental lack of understanding of a portion of the problem is uncovered (a working design solves *a* problem, but it must solve *the* desired problem). Understand the environment, the user, any software involved, and the use to which the design will be put.

2. **Think.** Discover several approaches to the solution. Do not stop at the first, most obvious solution [49, 50]. Keep attacking the problem, approaching in from different, nonobvious angles. Do *not* keep your mind in a straightjacket with concerns of the "correct" approach and what has been done before.

   With several approaches, consider: What are the advantages of each approach, disadvantages? Can they be combined to produce a better composite solution? Is a more general solution possible? Occasionally review the problem definition to ensure that the solution solves the problem completely. It may be the second, fourth, or tenth approach or composition of solutions that gives a solution which is both novel and practical and which is completely satisfactory [48] (see also Consider Unusual Techniques in Section 2.2).

3. **Design in Levels.** Only when you have the best solution should you begin to approach the implementation. The logic family, clock speed, and

other implementation-dependent details are postponed as long as possible. Algorithms, block diagrams, and flowcharts should be your tools first; gates and flip flops are used at the lowest levels. Use timing diagrams at all levels and let them define the circuitry — not the other way around. At each level, the design should be correct. There should be no loopholes or problems to be "attended to later." Only when the design at one level is correct enough to stand up to a design review should the next level be tackled [47]. See also Block Diagram and Timing Diagrams in Section 6.2 and Design Reviews in Section 5.2.

The documentation is *never* delayed until the project is finished but is a part of each design phase. Documentation, even if incomplete, can be of great assistance at all phases of the project. (See also Test Sequence in Section 5.1, Design Languages in Section 6.1, and Board I/O in Section 6.3).

 **High-Level Design.** When designing the architecture of a project, consider these inputs into the design process (adapted from Lincoln [2]):

- Review goals and results of similar projects.

- Understand thoroughly the environment in which the product is to be used. Consider the product's interaction with users or with other hardware or software.

- Project future tasks to be put to the product. Is the product general enough? Do you want to add extra features now or should they be postponed?

- Project marketability and the acceptable market price. What features does the market want? How large is the demand? How does the price impact the design?

- Evaluate current efforts in this direction, both in academia and by competitors.

- Be aware of the available and emerging technology.

- Appreciate the breadth of the demands to be placed on the project. For example, look at the constraints of the manufacturing, financial, thermal, electrical, mechanical, and reliability requirements.

- Solicit input from all quarters — hardware designers, programmers, potential users, mechanical engineers, manufacturing/testing personnel, field service, marketing and others. Do not wait until the project is complete to get this input.

As design begins, there are a number of tradeoffs to be made — the designer must find the fine line between extremes on a number of issues.

These tradeoffs are discussed in more detail later in this section; however, a few are summarized here:

- Hardware vs. software. Hardware has the advantage in speed, where software is usually cheaper and easier to develop and is cheaper to reproduce.

- Extremes of modularity. Too little modularity leads to a large, cumbersome result, but too much creates its own complexity with excessive interconnection between modules.

- Generality and completeness vs. orthogonal, sparse design. Do not omit useful or expected functions, but strive for an elegant, simple, straightforward interface.

**Compatibility.** When designing an upgrade, try to make it compatible with its predecessor. If it cannot be made 100% compatible from both the hardware and software points of view, the following should be done:

1. Document the anticipated differences early in the design so that there are no surprises.

2. After the project is complete, carefully document the differences.

This is a tricky issue. Incompatibility can be costly: old systems cannot be easily upgraded, new interfaces must be designed, supporting hardware and software cannot be used, and so forth. The interface to a system component such as a board should be as obvious and traditional to the programmer as possible (assuming that the highest level interface is at the software level). On the other hand, absolute hardware compatibility can seriously compromise the design. When compatibility cannot be absolutely guaranteed (perhaps for reasons of performance or space), consider a software driver on the processor to convert the user's customary input into the new form. This software interface can be a way to provide the best of compatibility and a superior design.

In designing a system, anticipate the future upgrade and try to minimize the compatibility headaches. One helpful procedure is simply a matter of documentation. Instead of documenting unused bits in a user-supplied value as "don't cares" (bits which have no effect on anything), specify that they must be 0. If, in the future, 1's in these bits are required to enable a new feature, we are assured that old users already have these functions disabled. Their software will run on the upgrade without surprises.

**Extensibility.** Think about future additions to the project. The design of these additions may be put off until later, but they should be considered during the high-level design. Without anticipating the additions, a designer may arbitrarily pick one of several alternate approaches, only to decide much later that the additions create a strong preference for one of the approaches *not* chosen [5, 6].

**Algorithm Choice.** In designing a module, consider the average engineer studying the design later. The engineer *must* be able to understand it quickly and correctly. If a clear technique must be replaced by a less-than-clear technique (perhaps for reasons of speed, cost, or board space), ensure that is is properly documented in the logics and in the textual documentation. Keep it simple, unless there is a very good reason to do otherwise.

**Minimize Coupling.** "Coupling" is a measure of the amount of interconnection between a module and its neighbors. Strive to minimize coupling. If the number of signals travelling from one module to its neighbors is excessive, this could suggest improper module partitioning. Try to make modules dependent on (receive signals from) as few other modules as possible [3].

A module with high coupling demands that anyone who wants to understand it must first understand something about all the modules with which it is coupled. Conversely, a module with low coupling puts fewer obstacles in the way of understanding and presents fewer opportunities for error. Not only is low coupling a benefit for understanding the design, but it is easier to design in an environment of low coupling.

**Information Hiding.** Do not let a module know *how* another module works. That is, only the output signals should travel from one module to another; internal signals should not be used. Output signals are those signals required by the *function* being performed, and internal signals are those suggested by the *algorithm* used to implement the function. In other words, a different algorithm might have different internal signals but would have identical output signals. If one module requires an internal, implementation-dependent signal from another module, this may be a clue that repartitioning is necessary.

Each module should be considered a black box. This allows a module to be replaced later with an equivalent module executing a different algorithm (perhaps faster or cheaper) with no penalty [4]. For example, a black box implementing multiplication could be executed with any of a large number of algorithms, each having different tradeoffs in cost, speed, accuracy, board space, and so forth.

**Generality, Completeness, and Symmetry.** Do not leave out any capabilities that are simple and reasonable extensions of the basic set of functions [5]. Ignoring this not only makes the set of functions difficult for the user to remember — with all its exceptions — but frustrates the user by suggesting functions that are not provided.

Symmetry suggests that similar functions be treated similarly. To use a processor's instruction set as an example, if the BRANCH-IF-POSITIVE instruction is able to use four addressing modes, the BRANCH-IF-NEGATIVE should use the same four modes.

The next two proverbs discuss the choice of commands by examining the opposing constraints.

**Maximize Orthogonality.** Highly orthogonal (literally, "right-angled") commands provide the most power. The geometric term "orthogonal" works well in conveying the impression of completeness without excess. An orthogonal set of vectors is adequate for accessing any point in its geometric space but is not redundant. It is necessary and sufficient. Extra vectors can be handy for more easily accessing important points, but too many of these become confusing and wasteful. Similarly, the minimum set of commands of a piece of hardware can be augmented with extra commands but too many become overly redundant.

The most extreme use of orthogonality in computer design would use a Turing machine as the processor; a Turing machine is a theoretical device that can simulate any computer but is much simpler than the simplest 8-bit microcomputer. This machine has no fat whatsoever. A more reasonable use of high orthogonality is seen in the Reduced Instruction Set Computers (RISCs), which have a different philosophy from the more mainstream Complex Instruction Set Computers (CISCs) [17]. The argument of the RISC proponents can be paraphrased: "Why implement this string search function or that complex addressing mode when each is used very infrequently? We prefer simple, fast computers that are easy to design, understand, and use."

The concept of orthogonality is pertinent to the choice of instructions in a computer instruction set and is also applicable to simpler groups of commands. Myers reports:

> [Orthogonality] is the objective of (1) holding the number of basic concepts to a reasonable minimum, (2) maximizing the independence among the concepts, and (3) avoiding superfluities. In other words, [a computer] architecture with low orthogonality has a large number of overlapping concepts, a number of "nice looking" but not particularly useful operations, and 17 different ways to zero or increment a register. An architecture with high orthogonality tends to provide more function at a given level of complexity or cost [6].

**Sparse Design.** "A designer has arrived at perfection not when there is no longer anything to add, but when there is no longer anything to take away" [8]. There is beauty in simplicity.

 **Conceptual Integrity.** Brooks said it best:

I will contend that conceptual integrity is *the* most important consideration in system design. It is better to have a system omit certain anomalous features and improvements, but to reflect one set of design ideas, than to have one that contains many good but independent and uncoordinated ideas. . . . Good features and ideas that do not integrate with a system's basic concepts are best left out. If there appear many such important but incompatible ideas, one scraps the whole system and starts again on an integrated system with different basic concepts [9].

 **Modularize.** Design in modules. Modules can contain any number of devices, but structured design dictates that each task must be divided into no more than a reasonable number of modules (perhaps three to eight). These modules themselves become tasks that are, in turn, subdivided. The divisions help in debugging, as well as design.

 **Hardware vs. Software Tradeoff.** A hardware implementation of an algorithm is faster, but software is more easily changed and is usually cheaper (both in design and manufacture). Software is a subset of hardware; some algorithms do not lend themselves to a software implementation. Consider the pros and cons of each during the high-level design to obtain the optimum tradeoff.

 **Robust Hardware.** Hardware should be "robust" — that is, forgiving and not damageable by the user. Anticipate the careless or malicious user and the harsh environment. For example, if the source on a bus is user-definable, either make that bus open-collector or open-emitter, or make sure that only one device can get onto a tri-state bus. Similarly, peripheral controllers must deny the user any chance to damage a mechanical device [5].

 **Interruptibility.** It is best if those functions that could be time consuming are interruptible (either aborted or put on hold). "Time consuming" might mean requiring half a second or more, although if the user is another piece of hardware, "time consuming" might be measured in microseconds. Users occasionally initiate a function that they decide to interrupt or stop; it is valuable to provide response as close to real time as possible.

**Processor Speed vs. Memory Size.** Amdahl's Constant says: "[A processor with a speed of] one megainstruction per second requires one megabyte of memory" [10]. Of course, applications vary, but this guideline is useful when deciding the amount of memory or the address space in a computer system.

**Memory Access Time vs. Instruction Time.** When designing a processor with a Harvard architecture, consider the ratio of the data memory access time to the average instruction execution time. (In a Harvard architecture, the memory containing data in a computer is different from that containing the instructions. A microcoded processor is an example because it uses different instruction and data memories. A graphics processor also has separate instruction and data (bit map) memories.)

A CPU executing one instruction per microsecond hardly needs a 50-nanosecond-access-time cache memory (the ratio here would be 0.05). This design would be compute bound: the memory sits idle most of the time waiting for the processor. On the other hand, a fast bipolar processor executing a quick and simple algorithm (such as copying data), which must communicate with a slow dynamic RAM or core RAM array across a system bus, might easily be an example of an I/O-bound process. Here, the processor is often waiting for the memory, and the memory-time-to-instruction-time ratio might be 4.0 or more. It is important to ensure that all elements of the architecture are fully used; idle hardware suggests wasted money. If all functions are compute bound, for example, the processor and memory are not properly matched, and a slower memory might be more economical (or a faster processor might be more powerful).

The definition of a "correct" memory-time-to-instruction-time ratio is, of course, highly dependent on the application and the architecture. The ratio does, however, provide a means of comparing two different computers performing the same algorithm. Computationally simple functions need a low ratio (1 to 3), while complex functions executing many instructions per data word can use a much higher ratio.

**Ergonomics.** The science of making the machine fit the user, rather than the other way around, is called ergonomics. Often, a board-level product does not affect the user and, hence, the designer of the product need not worry about the man-machine interface (MMI). Products that do interact with the user, especially those that generate video output, must give the MMI appropriate consideration. This can both enhance the value of the product and enable it to be sold to a wider market. West Germany has led the way by creating strict *laws* (not guidelines) governing the ergonomics of systems.

Other users, especially in Europe but also in the U.S., are taking ergonomics quite seriously, and the trend is toward increasing concern.

Regulated ergonomics deals almost exclusively with such physical issues as screen glare, contrast, flicker, color, angles of surfaces, distances between things, surface gloss, plus other issues of humans interacting with machines and with their work environment. Another side of ergonomics deals with the interaction of the human mind with the machine. These issues include response time, layout of a screen, meaningful but not extreme use of color, and input controls. It is important to maintain the user's interest. Similarly, considerable attention must be paid to the documentation. Poor documentation and an uninspiring or difficult interface are to the mind what poor terminal design is to the body [11, 12, 13]. (See also Fans in Section 3.2 and Safety in Section 5.3.)

**Allocation of Constants.** Be as generous as possible when deciding on unchangeable values. These values might be the maximum number of disks that can be controlled by a disk controller, the number of bits in a data word, the number of bits in an address, or the depth of a stack. Look for techniques that turn the constant into a variable. Instead of a fixed number of disk connections, could many disk drives be daisy-chained to a single controller? Instead of a fixed number of bits in an integer, could integers be stored as strings to allow almost unlimited integer range? Instead of a fixed address space, could virtual memory be used to extend the addressing range? Instead of a separate stack with a fixed depth, could the stack be put in main memory to give it a larger depth? Constant values cannot be changed and, if chosen with insufficient foresight, are confining [7].

## 2.2 Low-Level Philosophy

In Section 2.1, we have discussed some of the high-level topics in design philosophy. The basic approach developed, what are the lower-level philosophic concerns? We have not progressed enough in our design to be concerned with device specifications, but there are lower-level philosophic tools and ideas that are important.

**Technological Independence.** With memory and logic technology advancing very quickly, try to avoid placing constraints on the choice of devices. This allows an easier upgrade using newer parts at some future date [6]; an example would be anticipating the replacement of 64K dynamic RAMs (DRAMs) with 256K DRAMs.

**State Generators.** A state generator breaks up a time interval into many discrete parts. The use of a state generator instead of random logic to create periodic signals can be valuable — the signals are not dependent on one another and changing one signal does not modify the others. This approach is flexible, easy to understand, and easy to debug (see Figure 2.1).

Shift registers and counters make good state generators. The counter is more efficient: an n-bit counter has $2^n$ states, while an n-bit shift register with a single circulating 1 has only n states (see also Counter, Ring in the Glossary). However, the output of the shift register is more easily decoded. If a state is decoded from the outputs of a binary counter, be sure to *synchronize* before using it as a clock or to asynchronously set or reset (it may contain glitches). One technique gates the decoded signal with the clock. The clock effectively samples the signal after the output is stable to give a valid state. Shift register and Gray code counter outputs are typically immune to this problem.

**Gray Code.** A counter that outputs Gray code (instead of binary) numbers has the characteristic of changing one and only one bit each time it increments. (See the enumeration of the Gray code in Appendix 2.) This makes it more suitable than a binary counter for several applications:

- When a value is asynchronously sampled, it must be stable. Because many output bits might change when a binary counter is clocked, its value can be seriously misinterpreted when sampled (some bits might have changed before others). For example, in incrementing from the

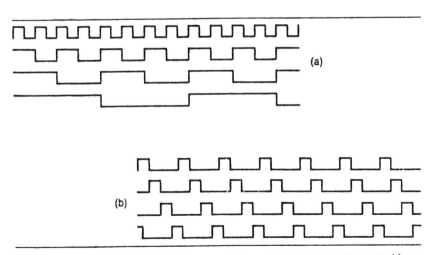

**Figure 2.1** State Generator Proverb: State generator outputs created from (a) counter and (b) shift register.

value 7 (binary 0111) to 8 (binary 1000), 4 bits are changing. A register sampling the value during its transition might load *any* of the 16 possible states of a 4-bit value. A Gray code counter changes only one output bit when incremented, eliminating sampling problems.

■ Because combinatorial functions of binary counter outputs can have glitches (caused by the fact that more than one output bit can change during an increment), these decoded functions must be synchronized before being used as an edge-sensitive input. The Gray code counter does not have this problem and is more suitable for use in a state generator.

Gray code counters are not widely available as off-the-shelf parts. Figure 2.2 shows the Gray code counting sequence as well as a simple technique for converting a binary value into its corresponding Gray code.

**Multiple Sampling.** When sampling a single asynchronous input, make sure that no more than one flip flop changes as a result. If two flip flops clock in the same signal, for example, they occasionally output different data. The causes can be one device having a slightly shorter setup time than the

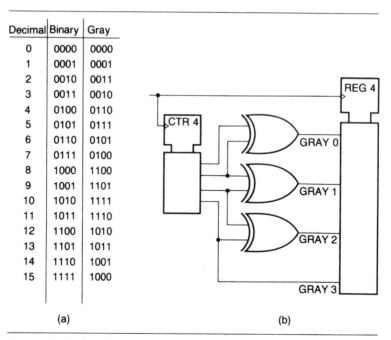

| Decimal | Binary | Gray |
|---------|--------|------|
| 0 | 0000 | 0000 |
| 1 | 0001 | 0001 |
| 2 | 0010 | 0011 |
| 3 | 0011 | 0010 |
| 4 | 0100 | 0110 |
| 5 | 0101 | 0111 |
| 6 | 0110 | 0101 |
| 7 | 0111 | 0100 |
| 8 | 1000 | 1100 |
| 9 | 1001 | 1101 |
| 10 | 1010 | 1111 |
| 11 | 1011 | 1110 |
| 12 | 1100 | 1010 |
| 13 | 1101 | 1011 |
| 14 | 1110 | 1001 |
| 15 | 1111 | 1000 |

(a)                                   (b)

**Figure 2.2** Gray Code Proverb: (a) Gray code vs. binary counting sequences and (b) Gray code output produced by a binary counter.

other or differences in clock and data line length. The solution, of course, is simply to eliminate one of the flip flops. Since it is commonly inadvertent, this situation can be difficult to notice.

**Board Communication.** Most boards communicate with the rest of the system via a few registers. Try to make them read/write instead of write only. Not only does this eliminate the need for the software to remember the contents of various registers, but it provides a way of verifying the data transfer to the board.

For easier debugging and board testing, look for any internal values to which a debugging engineer might want to have access. If a pipeline operates slowly enough or can be slowed, a "window register" might be valuable. With read-only access, the system processor can record the operation of the pipeline. Read-write access might be even better, allowing the engineer to inject test data into the pipeline. Look for other registers, memories, and so forth which you would like the processor to see. (See Debugging Philosophy in Section 5.1 for more information).

**Secrecy.** Consider ciphering for sensitive data that travels over insecure communication channels, is being stored in insecure memory, or is being stored on insecure media. Secrecy is usually important for military applications but is becoming increasingly important for commercial applications. New VLSI devices are making the encrypting and decrypting of information easier.

**Boundary Errors.** Watch for fencepost errors, off-by-one errors, and other boundary problems. (How many fenceposts do you need to make a 50-foot-long fence with fenceposts every 10 feet? The correct answer is 6, not 5, hence the name "fencepost error" for this type of confusion.)

A common boundary is found with comparisons. If we count occurrences of an event and want to announce the occurrence of the 23rd event, be careful. Do we want 23 completed events? the beginning of the 23rd event? the end of the 22nd? Approached from another angle, a comparator might output A < B for "not yet" and A > B for "go ahead," but which does A = B mean?

**Memory Hierarchy.** Often the price/performance ratio of memory can be optimized by well thought-out use of a memory hierarchy, consisting of high-speed registers, fast cache memory, medium-speed bulk memory (possibly using dynamic RAM devices), and slow disk storage.

**Writable Non-volatile Memory.** Aside from bubble memory (which is slow) and externally programmed devices (such as UVPROMs), there exist several different kinds of electrically-erasable PROMs (EEPROMs) and nonvolatile RAMs. These are random-access memories that can maintain their contents after power is turned off. Nonvolatile memory can be valuable for maintaining a small amount of data between operating sessions. Be careful, however, because these often need an unusual voltage for writing. Some nonvolatile RAMs have a limit on the number of write operations (this is called read-mostly memory). Consider also core memory and low-power RAMs with battery backup.

**Control Store.** Control store is the memory used to hold microcode. Consider writable control store (RAM) instead of permanent control store (ROM or PROM). This has several advantages: 1) the system can page in different routines that reduce the size requirement of on-board memory, and 2) modifying control store for debugging is fast. The disadvantages are: 1) extra registers and control signals are required to write to the RAMs and 2) the control store must be loaded after power-up before any processing can take place.

Loading the control store quickly may not be easy. Control store is usually not deep but is often many bytes wide. This is not a problem for many applications, however. Microcode for a new task is often loaded during a break, which allows use of a byte-wide or even 1-bit-wide serial path for loading.

**Sequential vs. Parallel.** Because MOS and TTL power consumption varies according to the internal switching, many devices operating in concert can put large noise spikes into the power and ground supplies over large areas of the board. MOS is typically quieter than TTL, but dynamic RAMs, for example, are serious offenders in large quantities. (For more information, see Logic Families in section 3.1). Decoupling capacitors and generous power and ground planes help, but there is also an architecture that might help. Consider many small sequential operations instead of a single large parallel operation. The large current spikes of many dynamic RAMs switching together, for example, can be reduced by serially accessing sections of the dynamic RAMs. A single access of a wide word in 200 nanoseconds might be replaced by four smaller accesses, one every 50 nanoseconds.

**Different Configurations.** Consider the effort to be expended by the user when one of several hardware configurations is to be selected. Software selectability is excellent when the user wants to use several configurations; however, if no more than one is used, a switch setting or jumper reduces the initialization demands. Either technique must be properly documented, whether the end-user or a company technician performs the configuration.

**Power-up Initialization.** Where applicable, the system should create an initialization pulse some time delay after power is applied (after everything is quiet and stable). Each module should consider where best to apply this pulse. The master clock must run during the initialization time to reset synchronous devices.

Avoid the situation in which one module is initialized and begins communicating with other modules before their initialization is complete. A solution is for each module to complete its initialization but not begin operation until either a certain delay has passed or another module has contacted it. The delay can be measured in seconds. Don't forget that during power-up, speed is measured according to human, not electronic, metrics (For more information, see Power Failure Detection in Section 2.3).

**Average Case vs. Worst Case.** Many algorithms require a variable amount of time to complete, depending on the inputs. For some applications, such as the implementation of a computer's instruction, this causes no problem. The speed of algorithms for these applications are judged against a continuous scale — the faster, the better.

Other applications are quite different. The creation of successive video images for a real-time display, for example, must be responsive to the frame-rate frequency (typically 60 Hz). A disk controller is controlled by the data rate from the disk. The external, immutable frequency gives these processes a fixed amount of time to do their jobs. The speed of an algorithm here is judged against a discrete scale: either the algorithm is fast enough *in its worst case* or it is unacceptable. An algorithm that is fast enough most of the time is little or no better than one that is always too slow. For these applications, the algorithm's worst case time is all that matters.

**Consider Unusual Techniques.** Always be aware of unusual techniques. Often, a difficult problem is only difficult because we are trying to perform the right operation with the wrong technique. A little imagination can go a long way.

Arithmetic, being a common task, has spawned a number of unusual computational approaches and representations. None is more widely used than other more traditional methods, but each has its own special benefits.

- **CORDIC algorithm.** The COordinate Rotation DIgital Computer (COR-DIC) uses a very simple algorithm. This algorithm has been expanded and can compute the functions of multiplication, division, sin, cos, tan, arctan, sinh, cosh, tanh, arctanh, en, exp, and square root. Each iteration increases the accuracy of the answer by one bit and requires three parallel adds or subtracts. This algorithm can be easily pipelined and, because it uses only simple arithmetic, is an excellent candidate for a hardware implementation [37, 38].

- **Continued fractions.** Positional notation (for example, decimal or binary representation) is the representation most commonly used by people and computers for recording numbers. An alternative notation represents numbers as continued fractions, which are numbers of the form

$$p_1 + \cfrac{q_1}{p_2 + \cfrac{q_2}{p_3 + \ldots}}$$

Each of the $p_i$'s and $q_i$'s are positional notation integers, but the entire value represents a real number. Some of the benefits of this notation are: faster multiply and divide than with positional notation, fast evaluation of trigonometric, logarithmic, and other unary functions, easy extension to infinite-precision arithmetic, infinite-precision representation of many important transcendental numbers, good provision for parallel hardware, and no roundoff or truncation errors [39].

- **Digit on-line arithmetic.** This technique performs operations on numbers one digit at a time (instead of all at once, as with typical computer arithmetic). It has many of the advantages of continued fractions without the need to represent the values with a foreign notation [40].

- **Modular arithmetic.** Instead of using a single word to hold a value, several smaller words can hold the value modulo several different moduli. (Remember that A mod B = the remainder after the integer divide A/B is performed. For example, 5 mod 7 = 12 mod 7 = 5.) Addition, subtraction, and multiplication can be performed on a value and another value in the same modular notation. The result, also in modular notation, can be converted back to binary. Because a single large word has been replaced with several smaller words that can be operated on in parallel without carries, a speedup is possible [41].

- **Power-of primes notation.** Logarithms make multiplication and division easy: two numbers are multiplied by adding their exponents and divided by subtracting them. By representing a number as the product of prime factors, the numbers are effectively converted into exponents. For example, $10 = 2^1 * 3^3 * 5^1 = \{1, 0, 1\}$ and $8 = 2^3 = \{3, 0, 0\}$. The product is computed $\{4, 0, 1\} = 2^4 * 5^1 = 80$ and 10/8 becomes $\{-2, 0, 1\} = 5/4$ [42].

- **Rational arithmetic.** Division is the only arithmetic operator that is not guaranteed to provide an integer result when given two integer operands. Division can force values out of the clean, simple, and accurate integer representation into the more complex and inaccurate floating point representation. Rational notation represents real numbers as a numerator/denominator pair (i.e., it postpones performing each division). Conversion into a floating point representation can be done with a single division at the end of a sequence of operations during which accuracy has been preserved [43].

- **Bit vector machines.** Extremely wide data words (hundreds to thousands of bits) are used in bit vector machines. These words are combined with logical operations only (no arithmetic), but the massive parallelism of the wide word can lead to significant speedups in arithmetic operations [42].

- **Incremental methods.** Some applications require solving a mathematical expression at many sequential points. Instead of calculating the value at each point from the expression, incremental methods use the fact that the expression has been solved for the point's predecessor. One type of incremental method, forward differences, works well for polynomials. After initialization, the value of the equation at each point is calculated with n simultaneous adds based on the previous value [16, 44].

**Shifting for Fast Divide.** Since shifting is very fast with hardware, some algorithms replace division by powers of two with right-shifting. Note, however, that these two operations are not exactly identical. Since the result of an integer division may not be an integer, a question must be raised about the relationship between the integer quotient produced and the *actual* value (i.e., how do you handle the fractional part of the answer?). With two's-complement numbers, divide and right-shift produce the same answers for positive numbers, but produce different values for negative numbers. The problem is that divide is usually defined to round toward zero ($11/2 = 5$ and $-11/2 = -5$) but shift right rounds toward negative infinity (RSHIFT{11} = 5 and RSHIFT{-11} = -6). One correction is to conditionally add 1 to a negative result, based on the shiftout value. Alternatively, the hardware can negate, shift, and negate. The main thing to keep in mind is simply that a right-shift is not equivalent in all cases to division by two [14].

## 2.3 Reliability and Testability

Logic design may be thought of as remote from the imprecision of analog design, yet digital logic is in fact a subset of analog logic. Noise and error are often present; this section presents some tactics we can use to counter them.

**Self-Checking.** A desirable feature is the ability to perform self-checking. A power-up confidence test is therefore worth some extra hardware in some applications.

Diagnostic software or hardware can also be invoked during idle periods. This helps identify problems as soon as possible. However, ensure that the time required by the diagnostics does not interfere with the execution of the system.

Self-checking does not eliminate hardware errors but it does reduce the damage they cause. An error with no explanation is much worse than an error with information provided to the operator explaining the error and recovery procedures. (For more information, see Status Displays in Section 4.2.)

**Testability.** During design, consider the testing of both the prototype board and the production boards and how it can be made easier. Expect the worst and plan for fixing it.

**Reset Often.** If a device is periodically reset or two circuits are resynchronized, try to make the resetting or resynchronization happen as often as possible. For example, a shift register can be used as part of a state generator by looping the shift-out data into the shift-in input and initializing the shift register with a one in a single bit position. Without periodic resetting, however, this circuit remains faulty after an error. Soft errors (caused by power supply spikes, static, electrical noise, radiation, test probes, human fingers, and so forth) should be expected and must be corrected as often as possible. This may not help errors in instruction streams — one error here is often fatal — but many situations are fault tolerant and do not mind infrequent errors if the errors' effects are minimized. Try to prevent a soft error from causing a permanent, instead of transient, logic error [18]. (For more information, see State Generators in Section 2.2 and Error Sources in Section 2.3.)

**Power Failure Detection.** Some power supplies produce a digital signal that indicates power is available for only a short time longer (perhaps several milliseconds). An analog-to-digital converter sampling the power input can optionally be added to provide an indication of the health of the supply. When this information is or could be available from the power supply, consider installing a power-down sequence that, when invoked, would immediately preserve data or safeguard hardware. This sequence could retract the heads of a disk drive or save data in nonvolatile storage, for example [19].

**Anticipate Metastability.** If metastable conditions are unavoidable, calculate the average rate of metastable occurrences to see if it is acceptably low [27]. There are ways of avoiding or minimizing metastable problems (see Metastability, Section 2.4).

**Error Sources.** Errors can be classed as hard (permanent) or soft (transient). Hard errors are permanent device failures and have several main sources:

- **Quality.** Parts of marginal quality tend to fail early.

- **Stress.** Excessive stress can occur at any time in a part's life. The stress can be excessive voltage on inputs, heat or cold, mechanical forces, or nuclear radiation. Derating is often used to reduce the typical demands on the device and, therefore, reduce the occurrence of excessive stress.

- **Wearout.** As a part ages, it can become prone to failure.

These failure mechanisms combine to form the bathtub curve, which predicts the composite failure rate (see Figure 2.3).

Soft errors are less disasterous; often, operation can resume after the transient effects are gone. However, because they are transient, their sources can be much more difficult to pinpoint and correct. Some sources are:

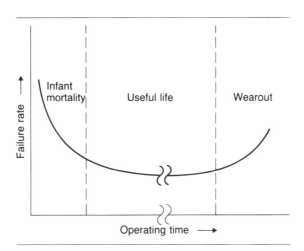

**Figure 2.3** Error Sources Proverb: Bathtub curve.

- **Radiation.** Dynamic RAMs use the charge in a very small capacitor to store a 0 or a 1. The charge is so small that alpha particles (either from external sources such as cosmic rays or from radioactive impurities in the device itself) can alter the value.

- **Power/ground glitches.** Poor decoupling or poor power or ground distribution on the circuit board can cause a noisy power or ground supply to chip. When a chip's reference voltages change, the chip's definition of 0 and 1 inputs alters, possibly causing an error.

- **Metastability.** When the data setup and hold times are not met in a flip flop or other synchronous device, this temporary nonlogical state can occur (see Metastability in Section 2.4).

- **EMI/RFI and noise.** Signals can be altered by the radiation output from nearby wires (crosstalk) or by sources external to the system (see Radiation Emission in Section 3.2).

- **Nuclear radiation.** Severe radiation can cause transient logic failures. Solutions are shielding and the use of radiation-hard logic families.

Some of these problems have solutions that reduce or prevent occurrence of the problem. Other solutions allow operation in spite of the problem. These solutions are discussed in this and later sections.

**Error Handling.** Possible responses to an error are: continue and ignore the error, continue but report the error, halt, halt and report, retry, reset the failed subsection, and use error correction to eliminate the effects of the error. Consider the response that best suits the application.

**Memory Scrubbing.** This technique improves the reliability of dynamic RAM memory systems. The dynamic RAM chip refresh cycle, which typically occurs over 60,000 times per second, can be used to not only refresh the chips, but read the data and correct it if necessary. Since the RAMs must be accessed anyway, this is an ideal opportunity to correct any errors.

A DRAM is given its address in two separate loads; the two halves of the address are the row address and the column address. Chip refreshing is performed by accessing each row address periodically. The column address, which is not important for chip refresh, can be used to sequentially sample all the values in the row. After the row address cycles through its states, the column address is incremented before the next sequence of row addresses. Using 64K x 1 dynamic RAMs, this technique samples every word in the entire memory roughly once per second and does not degrade performance.

Correcting single-bit errors as often as possible reduces the likelihood of their becoming two-bit errors, which are uncorrectable with the most common ECC schemes [35].

 **Consider Fault-Tolerance.** Important areas of fault tolerance [46] are:

- **Serial Data Stream Error Detection/Correction.** Consider CRC (Cyclic Redundancy Check) for detection and RS (Reed-Solomon) codes for burst error detection and correction. Parity appends a bit to a data word, which records whether the word contains an odd or even number of 1 bits. The parity can be recomputed at any time to see if the parity bit remains valid on if it has changed due to an error. These techniques are valuable for communication, peripheral I/O, and other serial data streams. The serial streams used here are one bit wide and arbitrarily long.

  When CRC is used, a checksum appended to the message is compared with that computed by the receiving device. If they differ, an error has corrupted the message. Such an error might be correctable if RS codes are used, but error correcting codes require redundancy (that is, the message sent is longer than the original message).

- **Parallel Data Error Detection/Correction.** Consider parity for detection and error-correcting codes (ECC) for detection and correction. These are valuable for parallel data, such as that from memory or on a bus. A one megaword memory with 16 bits per word made from 64K dynamic RAMs (DRAMs) with parity has a mean time between failures (MTBF) of 0.8 years, while one with 6 bits of ECC per word has an MTBF of 8.2 years [21]. Memory scrubbing can further improve reliability by correcting 1-bit errors before they become 2-bit, uncorrectable, errors. ECC can even compensate for the complete failure of one 64K x 1 DRAM chip.

  Even parity detects the all-ones failure on data words of even sizes; odd parity detects this error on odd-sized data words. An all-ones error causes all lines on a bus to be always true. Only odd parity detects the all-zeros failure. Several different schemes exist for improving error detection by adding extra parity bits.

  All ECC schemes require some extra bits to provide the redundancy to detect and correct errors. A "(p, q) code" requires p bits to encode a q-bit value. The BCH code is a $(2^m - 1, 2^m - 1 - mt)$ code, where t is the number of errors to correct. For example, it can be a (31, 21) code and correct up to two errors in each 31-bit word. Each 31-bit word can be decoded into 21 data bits. A (127, 91) code corrects up to 9 errors in each 127-bit word. With t = 1, the BCH code becomes the commonly-used Hamming code [22].

■ **Module-Level Redundancy.** Modules or boards can have fault detection circuitry that enables a backup module when an error is detected. This, of course, can be several times more expensive than the equivalent system without this redundancy. In some applications, however, the extra cost is quite acceptable when compared to the cost of failure; these applications include spacecraft computers and high-reliability telephone systems.

**Signature Analysis.** This valuable testing technique works in a manner similar to that of a CRC check: a "signature" for a serial stream is computed and compared with the known correct signature for that stream. If the signatures differ, that part of the logic which generated the stream must be faulty. This technique can be used both to test the board *and* to hunt down the broken section. A signature analyzer can be used manually, as a separate piece of test equipment, or the analyzer can be incorporated into the board. If the stream is very high frequency data, the on-board method may be the only one possible. Figure 2.4 outlines the design of a signature analyzer.

Incorporated into the design, the analyzer consumes perhaps 5 to 15 chips (depending on clock speed, number of signal streams to be analyzed, and the additions required to the bus interface). This is certainly the most automatic way to use the technique; software on the system processor can check the many signatures internal to the board and report in a short time either 1) that the board is working or 2) that it is not working, plus the approximate location

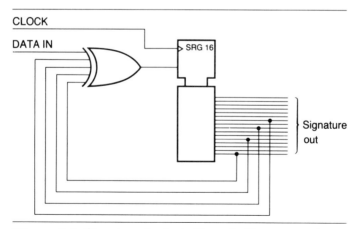

**Figure 2.4** Signature Analysis Proverb: Signature analyzer. The serial data in is converted into a 16-bit signature.

of the error. This is thoroughly documented elsewhere [20, 45], but several points can be summarized here:

- Signature analysis is excellent for automatic testing and confidence tests. It compresses long data streams into a single, compact, easily-understood value.

- It is especially good for testing production boards but can also assist prototype debugging (to give a quick regression test to areas of the board previously debugged).

- The same signature can be obtained many times; if the signatures are not all identical, a transient error is indicated.

- Many video tests are visual, requiring the attention of a test technician. Signatures can test video and memory streams to provide automatic tests to replace visual ones.

- If the signature analyzer is off-board, a thorough debugging sequence should be developed, defining test points and correct signatures to be expected. The software to exercise an on-board analyzer must also be carefully planned to obtain the most benefit. Try to automate the documentation of the correct signature for each signal. A single wire change can easily change several signatures, complicating manual documentation.

- Start and stop times for analysis must be defined. Each signal being sampled must be used to create a signature only during a certain time window.

- For an on-board analyzer, the user must send the signal number to be analyzed. This number forms the address to a multiplexer, which picks the signal.

- Analysis can occur periodically during operation (if possible) for a confidence check. Alternatively, it might happen only after user initiation.

The bit streams described so far have been data from a single source over time. An alternative looks at the data from many sources at a single time instant. This alternative can be implemented in many situations by replacing registers with shift registers. Shift registers are cascaded, connecting the shift-out bit into the shift-in of its neighbor, to form one large datum representing the state. During normal operation, the shift registers operate like the registers they replace. To form a signature from the state, operation is halted and the shift registers shift the data into the analyzer. This type of analysis is not as beneficial in isolating the source of a problem and can not run concurrent with normal operation. Depending on the application, however, its conversion of a large part of the machine's state into a single number may have advantages.

**Dead-Man Timer.** Also called a "watchdog timer," this is a physical timer that resets the board or board subset if the timer is not periodically restarted [19, 23]. Valuable for both software and hardware processes, the timer is periodically restarted by a correctly functioning process; when something goes wrong, the timer hopefully "times out" and resets the system to a known state. A commonly-used example is bus time-out [24], in which a device on a bus is removed by the bus master after taking too long to perform a task.

Systems used to debug low-level software often make use of this technique by frequently restarting a hardware timer. If the software goes crazy, perhaps executing data or remaining in a subroutine too long, the timer restarting is not done, the system is reset, and the processor resumes execution at a known state.

 **Service Frequency.**

One of the rules of thumb upon which electronics test engineers seem to agree is that the average printed circuit board will malfunction and require service *two times* over its lifetime. Although this rule of thumb is undoubtedly high when applied to small, simple boards, it is also undoubtedly low for large boards, and for boards that will be subjected to a harsh environment (e.g., static discharge on inputs, or high temperature, or high humidity) [28].

This underscores the importance of creating a reliable, well-documented, easily-tested board.

## 2.4 Low-Level Logic Considerations

While trying to remain aloof from the details of logic families, let's consider some general issues concerning the physical nature of design as well as other low-level concerns. Several references to actual logic families or devices have crept in. Section 3.1 discusses the families in detail.

**Basic Dos and Don'ts.** A few of the fundamentals of low-level logic design are:

- *Do* examine worst-case timing for all devices. Look out for race conditions, in which two combinatorial input signals are effectively racing. In this situation, the output is not only a function of the input signals, but of which came first. While worst-case timing usually implies the maximum delay through a device, it can instead be the minimum delay. If no minimum delay appears in the manufacturer's data book, assume it to be zero nanoseconds (true, a propagation delay is never this fast, but plan for the extremes). In short, the worst-case timing analysis examines whatever is worst for your design.

- *Do* use synchronous design techniques.

- *Don't* use pulses formed by device delays. Pulses should only be formed by synchronous devices.

- *Don't* use cute tricks which someone else would have difficulty in understanding. The design must be maintainable.

**Propagation Delay Calculation.** If a data input to a synchronous device (flip-flop, counter, register, shift register, and so forth) is the output of another synchronous device, the propagation delay is easily calculated. If the input comes from a combinatorial device (gate, adder, ALU, comparator, multiplexer, demultiplexer, memory element, and so forth), trace the input back to its *synchronous* origin to find the worst-case delay. Also, consider the additional effective delay caused by these factors:

- **Different Clocks.** Watch out for clocking two successive pipeline stages with different clocks. Often a single master clock signal is fed into several gates to create clocks which together can drive more loads. However, the different delays through these gates and the different path lengths of the clock signals means that these new clocks are phase shifted with respect to each other. That is, the edges do not line up. The clock period can effectively be smaller when two successive synchronous devices are clocked with different clocks.

- **Capacitance.** Always look at the capacitance for which a propagation delay is guaranteed by the manufacturer. Input capacitances of devices plus signal capacitances can add up quickly (see Signal Capacitance in Section 3.3). Signals in the 4000 CMOS family are delayed by 1 nanosecond (ns) with as little as 1 picofarad (pF) of additional line capacitance. 74S TTL is delayed by 1 ns with each 25 pF of additional capacitance (less delay for Advanced Schottky, more for Low-power Schottky). ECL rise/fall times are also affected by capacitance but the delays are also dependent on the line termination, which is selected by the designer. ECL is roughly as sensitive to capacitance as 74S TTL [25, 34].

- **Path length.** The signal is delayed an amount proportional to the length of the wire. The best transmission media propagate a signal at slightly over 1 ns per foot. In typical applications, the rate is 1.5 ns or more per foot.

- **Ringing.** The settling time (the time for the signal at the input to stabilize) may be significant. If ringing is so bad that a valid, stable signal is not present at the destination until the wavefront has reflected up and down the wire several times, the path length delay has effectively become several times greater.

As the logics are sketched out, check every off-page input signal to find the propagation delay already on it. Make sure that the additional delay added to the signal by the current page is not excessive.

**Hold Time.** Watch for hold times on data *and control* inputs. Often, signals used as inputs to a synchronous device must be stable both before the clock edge (setup time) *and* after (hold time). For example, picture the output of a 74S74 flip flop feeding the mode input of a 74S299 8-bit shift register (see Figure 2.5). If the two devices are clocked with the same clock (at first glance, not an unreasonable design), there will be a problem. The mode input requires a *5 ns hold time*, but the minimum hold time of the 74S74 output is unspecified.

**Violating Worst Case Timing.** In rare cases (hopefully they are very rare), you may find that you have no choice but to violate worst-case timing. That is, the worst-case propagation delay between two points is longer than proper synchronous design demands. Tolerating such a violation is a risky undertaking. It creates, in essence, an imaginary specification for certain parts with performance requirements more stringent than the manufacturer guarantees. Make sure that there is a very good reason for doing this by examining the possible alternatives (are the alternatives too expensive? too slow? do they consume too many chip locations?). Document all signals that do not

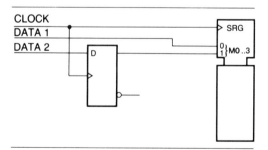

**Figure 2.5** Hold Time Proverb: Since the flip flop and the shift register are clocked with the same clock, we have a problem: the flip flop is not guaranteed to provide the 5 ns input hold time for the shift register.

meet worst-case timing in the logic diagram on which the signal appears and in the textual documentation. Be sure that the debugging/testing document also references this situation.

Some manufacturers sell, at a premium, parts with superior specifications. Consider also the feasibility of a separate IC test fixture that selects the higher-performance chips required [15].

**Metastability.** More than one clock crystal in a system implies the possibility of metastability, as does the use of asynchronous communication. Metastability can occur if the setup time of a synchronous device is violated. The setup time is violated if a data or control input does not arrive in enough time before the clock. The output of a device in metastability can be in a nonlogical state (some voltage between low and high) or the device can be unresponsive to clock inputs for many propagation delay times [26]. In fact, the maximum duration of the metastable condition cannot be defined, but the duration could in practice be *1000 times* the normal response time [27]. This problem has been known for over 20 years and yet is probably the cause of many transient problems in "debugged" digital systems.

An example of a dangerous situation is the sampling of a 6 MHz asynchronous data stream with a 10 MHz clock using a 74S74 flip flop. The flip flop becomes metastable *600 times per second* [25, 27]. Several remedies exist; the first prevents metastability from occurring, while the others try to minimize its effect:

- Use only one clock in a system, do not communicate asynchronously, and design completely synchronously.

- Use the fastest possible parts to synchronize asynchronous signals.

- Use the newer fabrication processes, such as Fairchild's FAST and Advanced Micro Devices' IMOX, which are more immune to metastability [25, 27].

- Use fault-tolerant hardware (see Section 2.3).

- Synchronize asynchronous signals through two levels of flip flops (see Figure 2.6). Even if the first flip flop becomes metastable, the second hopefully outputs only valid, logical levels. Note that this delays the signal.

**Programmable Logic.** Programmable logic includes Integrated Fuse Logic (IFL), Programmable Array Logic (PAL) and Programmable Logic Arrays (PLAs). These can be valuable for replacing large amounts of logic. Often, a correction or upgrade can be painlessly implemented with the replacement

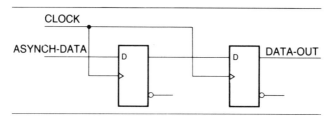

**Figure 2.6** Metastability Proverb: Two levels of flip flops to minimize the effects of a metastable condition. If the first flip flop becomes metastable, hopefully it recovers before sampling by the second flip flop.

of one of these chips. Devices from the various manufacturers provide programmable AND arrays, programmable OR arrays, flip flops, and feedback of one of these chips. Devices from the various manufacturers provide programmable AND arrays, programmable OR arrays, flip flops, and feedback

**Use of PROMs and Programmable Logic.** PROMs can be useful in many logic applications (as opposed to memory applications), such as decoding, black-box logic functions, and so forth. For example, a PROM could be used to implement a state machine. The old state plus several data values could form a PROM address; given this address, the PROM would output the new state. Remember, however, that PROMs and programmable logic (PLAs, IFL, and so forth) may pose manufacturing difficulties:

- They must be programmed and, therefore, cannot be used off-the-shelf (ROMs are an exception, of course).

- They may use a logic family that is more static-sensitive than TTL, which can imply bothersome precautions or a high percentage of failed parts.

- Second sources do not always exist (second sources are manufacturers that produce parts identical to those made by the original manufacturer).

- Since their contents are variable, they must be well documented.

This list is not intended to discourage the use of these valuable devices but to emphasize the ways in which they differ from other devices. (For more information, see Programmable Device Documentation in Section 4.2.)

**Programmable vs. Semi-Custom vs. Custom Logic.** A number of approaches to user-defined logic exist, each with its own merits in initial cost, per-unit cost, generality, and time required. Programmable logic and

PROMs have minimal initial cost, are somewhat general, and require very little time to customize (measured in hours).

When space is a premium and the sales volume supports it, more expensive techniques are used. Semicustom and custom logic facilities take a customer's design, perhaps implemented with 5 – 100 SSI and MSI devices, and produce custom integrated circuits performing the same function. The new devices usually occupy less board areas, consume less power, and are less expensive than the parts they replace. Semicustom techniques such as uncommitted logic arrays, gate arrays, and standard cells require a small number of months and a significant financial investment but are cheaper than the parts they replace when thousands of units are created. Full custom logic requires even more time and money to develop, but provides the highest density and is the choice when tens of thousands of units are built.

**Power Consumption with MOS.** Since MOS consumes significant amounts of power only when signals switch states (i.e., change from high to low or from low to high), try to minimize switching if possible. Switching can be minimized by disabling the clock to a device when it need not change, or using a lower frequency clock for certain sections. This becomes important in MOS designs as well as TTL designs considering the 74HC TTL-compatible CMOS family.

While 74HC consumes only marginally less power then 74LS at the higher frequencies (10 MHz and up), remember that not all gates switch state after each clock. For example, the least-significant output of a 4-bit counter with a 10 MHz clock changes at a 10 MHz rate. However, the other outputs will change at $1/2$, $1/4$, and $1/8$ the 10 MHz rate. When switching at frequencies lower than 1 MHz, 74HC consumes much less power than 74LS. (For more information, see MOS Comparison and Logic Families in Section 3.1.)

**Use LSI.** High density LSI and VLSI devices reduce chip count and provide debugged circuit modules. In comparing the cost of a circuit implemented with SSI/MSI chips vs. its LSI equivalent, include the costs of board space, chip insertion, and testing in addition to the chip costs. On the negative side, LSI can be less reliable and second-source manufacturers may be hard to find.

**Beware of New VLSI.** Complex new chips, like microcomputers, take time to perfect. However, with the marketplace competition as aggressive as it is, chip manufacturers often rush less-than-perfect parts into production. Second sources for new devices are often not planned. Avoid being a guinea pig, if possible.

**Pullups for Sets and Resets.** Use separate pullup resistors for flip flop sets and resets. A flip-flop can then be set or reset by grounding an input. Put as few signals on a single pullup as possible. This allows a technician or an automatic board tester to set or reset a flip flop, for example, and minimally affect the rest of the board. (For more information, see TTL Inputs Tied High in Section 3.1 and Pullups in Section 3.2.)

**Clock Edges.** It is best to use only one edge for clocking synchronous devices. Mixing rising-edge-triggered and falling-edge-triggered devices can cause confusion. Rising-edge-triggered devices are the most popular; it is probably best to use them.

**Master Clock.** Where possible, use the master clock to clock a synchronous device instead of an arbitrary signal. It is cleaner not to qualify (conditionally disable) the clock with a gate but rather to use a mode signal to *internally* qualify the clock. This is done on shift registers and counters which have a HOLD/ENABLE input (with HOLD selected, incoming clocks have no effect). The common 74374 8-bit register does not have this feature, but the Am29825 8-bit TTL register and the 10131 2-bit ECL register both have a clock enable.

**Asynchronous Functions.** Be careful when using the asynchronous functions of a device. Occasionally, you have a choice between two devices, one of which has a synchronous function and the other, an equivalent asynchronous function. An example of such a pair are the 74161 counter (with asynchronous clear) and the 74163 counter (with synchronous clear).

Sets and resets on flip flops are also asynchronous. Synchronous control lines have setup and hold requirements, but asynchronous lines do not. Since these asynchronous control lines can affect the states of synchronous devices as a clock does, they must be thought out as carefully and made as synchronous as other control lines. Of course, one of the basic tenets of synchronous design is that data lines *are* allowed to glitch during transition.

**Registers vs. Latches.** Do not use a latch where a register will work. Registers use a clock edge to sample the input at a single instant, while latches continuously allow the data through during one phase of the clock. The register is more synchronous and allows transients on the signal to die out before the signal is used.

**Noise Immunity.** Both TTL and ECL logic are more sensitive to noise in low-level signals than those close to ground. For this reason, keep signals high as often as possible and let them drop to indicate a data pulse. In other words, look at the duty cycle and create the signals to be high most of the time. This is important for long or noisy lines, such as those in busses or backplanes [29] (see Logic Families in Section 3.1).

**Tri-State Bus.** When no device is enabled to a tri-state bus, TTL bus signals float to about 2 volts. This nonlogical state can be confusing during debugging. Consider having something on the bus at all times, especially on cyclic busses (such as the RAM address bus on a memory board).

Do try, however, to put in a "dead band" between two successive bus users. This band is a gap during which no source drives the bus. Tri-state drivers are stressed when more than one is active at a time (see Tri-State Outputs in Section 3.1).

**Logic Simplification Techniques.** Considerable emphasis has been placed on logic simplification/minimization techniques in the past. Given a Boolean equation representing a combinatorial circuit, these techniques produce an optimal equation which translates into a circuit using a minimum of gates. Such emphasis is somewhat misplaced today [30] for several reasons:

- We now have valuable functions such as PROMs, programmable logic and (most importantly) microprocessors. Logic families also provide a greatly increased choice of MSI and LSI devices. Before this became available, designers had little to work with other than gates; their use of considerable SSI made simplification techniques valuable tools.

- With debugging and testing costs an ever-increasing multiple of chip cost, gates are now used in simple and intuitive ways. Instead of squeezing out every possible gate in a combinatorial circuit, functions are often left in their original, easy-to-understand form. This helps to increase maintainability.

Although less important, these techniques have their applications. Some of the minimization techniques available are:

- **Algebra.** The rules of Boolean algebra can be used to simplify logic when a function is written as an equation (see Simplification Techniques in Section 1.1).

- **Karnaugh maps (Veitch diagrams).** Perhaps the most widely used method, Karnaugh maps are drawn with a rectangular grid. The states of some variables are written in Gray code across the top and the state of the remaining variables are written along the left side. Each cell of the grid describes a possible state of the inputs and holds the desired output truth value. This method is more fully described in Simplification Techniques in Section 1.1 [31, 36].

- **Star algorithm.** Developed by Roth, this algorithm lends itself to computer implementation. An n-dimensional cube (where n is the number of variables) contains the possible states of the input variables at its vertices [32, 33].

## 2.5 Helpful Circuits

A few helpful, nontrivial circuits or circuit ideas are contained in this section. When you discover a new one, make a note of it here.

**Clock Distribution.** A high-frequency clock might be distributed to many loads across a noisy medium (like a backplane) with the following technique:

1. Send out a medium-frequency (10 MHz, for example) digital clock from the source.

2. Receive the signal with a high-impedance receiver to accommodate many receivers.

3. Use a bandpass filter to extract the desired harmonic (20 MHz, 40 MHz, etc.). This filter tuned to the desired frequency ensures a reliable on-board clock.

4. (optional) Pass the signal through a variable delay line to shift the clock in phase with clocks on other boards.

5. Clean up the signal with a Schmitt trigger, preceded by a conversion from an analog voltage to a digital voltage.

Consider putting the delay line and a clock tap at the edge of the board so that the clock can be phase-shifted while the system is running.

**Static RAM Writes.** For a high-speed write of a block of data into a static RAM, try holding the WRITE line continuously active while changing the address lines. The data must change *after* the address has changed. Use a Gray code for the address sequencing so that only one bit changes at a time. This works only if there are no internal decoding spikes on the address. (Consider this application carefully before using it!)

## Level-to-Pulse Converter.

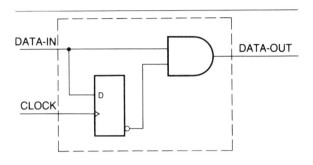

**Figure 2.7** Level-to-Pulse Converter Proverb:
Level-to-pulse converter (a) and timing diagram (b).

 ## Pulse-to-Level Converter.

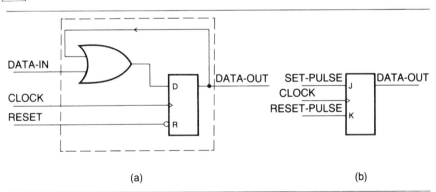

(a)                                              (b)

**Figure 2.8** Pulse-to-Level Converter Proverb: Pulse-to-level converter made
with a D flip flop (a) and J-K flip flop (b).

## References

[ 1] Brooks, F. P., *The Mythical Man-Month* (Addison-Wesley, 1975), p. 180.

[ 2] Lincoln, N. R., "Supercomputers = Colossal Computations + Enormous Expectations + Renowned Risk," *Computer*, May 1983, p. 38.

[ 3] Yourdon, E. and Constantine, L. L., *Structured Design* (Prentice-Hall, 1979), pp. 84–104.

[ 4] Ibid., pp. 252–253.

[ 5] Welsh, R. L., "Qualities of Good Software," *MIS Week*, October 13, 1982, p. 38.

[ 6] Myers, G. J., *Advances in Computer Architecture* (John Wiley and Sons, 1978), p. 295.

[ 7] Bentley, J. L., "Programming Pearls," *Communications of the ACM*, January 1984, p. 12.

[ 8] Bentley, J. L., "Cracking the Oyster," *Communications of the ACM*, August 1983, pp. 551–552.

[ 9] Brooks, pp. 42 and 46.

[10] *The Computer Museum Report*, Winter/1983 (Marlboro, MA: The Computer Museum, 1982).

[11] Manuel, T., "Molding Computer Terminals to Human Needs," *Electronics*, June 30, 1982, pp. 97–108.

[12] Knapp, J. M., "The Ergonomic Millennium," *Computer Graphics World*, June 1983, pp. 86–93.

[13] Snyder, H. L., "Visual Ergonomics and VDT Standards," *Digital Design*, February 1983, pp. 24–34.

[14] Steele, G. L., "Arithmetic Shifting Considered Harmful," (Cambridge, MA: MIT AI Memo # 378), September 1976.

[15] "Careful Screening Makes Chip Run Four Times Faster," *Integrated Circuits Magazine*, January/February 1984, pp. 63–65.

[16] Knuth, D. E., *The Art of Computer Programming*, vol. 2 (Addison-Wesley, 1969), p. 431.

[17] Patterson, D. A. and Piepho, R. S., "RISC Assessment: A High-Level Language Experiment," *Ninth Annual Symposium on Computer Architecture*, April 1982, pp. 3–8.

[18] Blakeslee, T. R., *Digital Design with Standard MSI and LSI* (John Wiley and Sons, 1979), p. 146.

[19] Oppenheimer, C. P., "Reliable Designs Begin with the Basics," *Computer Design*, August 1983, pp. 93–99.

[20] Peatman, J. B., *Digital Hardware Design* (McGraw Hill, 1980), pp. 162–172.

[21] *Memory Databook 1983*, Second Edition (Santa Clara, CA: OKI Semiconductor, 1983), pp. 250–257.

[22] Siewiorek, D. P. and Swarz, R. S., *The Theory and Practice of Reliable System Design* (Digital Press, 1982), pp. 89–90.

[23] Ibid., pp. 110–112.

[24] Blakeslee, p. 254.

[25] *FAST Circuits and Applications Seminar* (Mountain View, CA; Fairchild Corp., 1982).

[26] Chaney, T. J., and Molnar, C. E., "Anomalous Behavior of Synchronizer and Arbiter Circuits," *IEEE Transactions on Computers*, April 1973, pp. 421–422.

[27] *Bipolar Microprocessor Logic and Interface Datalook* (Sunnyvale, CA: Advanced Micro Devices, Inc., 1985), pp. 8–2 – 8–3.

[28] Peatman, pp. 146–147.

[29] Blood, W. R. et al., *MECL System Design Handbook*, Revised Edition (Phoenix, AZ: Motorola Semiconductor Products Inc., 1980), pp. 193–194.

[30] Peatman, pp. 45–49.

[31] Winkel, D., and Prosser, F., *The Art of Digital Design* (Prentice-Hall, 1980), pp. 22–29.

[32] Roth, J. P., *Computer Logic, Testing and Verification* (Computer Science Press, 1980).

[33] McCoy, E. G., "The ABCs of Simplifying Logic Equations, Simply," *Computer Design*, July 1982, pp. 99–102.

[34] *F100K ECL User's Handbook* (Mountain View, CA: Fairchild Corp., 1982), pp. 4–10 – 4–12.

[35] Madewell, J., "DRAM Controller Helps Scrub Errors," *Electronic Products*, September 17, 1984, pp. 88–92.

[36] Blakeslee, p. 40.

[37] Ahmed, H. M. et al., "Highly Concurrent Computing Structures for Matrix Arithmetic and Signal Processing," *Computer*, January 1982, pp. 65–80.

[38] Walther, J. S., "A Unified Algorithm for Elementary Functions," *Proceedings of AFIPS 1971 Spring Joint Computer Conference*, 1971, pp. 379–385.

[39] Seidensticker, R. B., "Continued Fractions for High-Speed and High-Accuracy Computer Arithmetic," *Proceedings of the 6th Symposium on Computer Arithmetic*, June 1983, pp. 184–193.

[40] Watanuki, O. and Ercegovac, M. D., "Floating-Point On-Line Arithmetic: Algorithms," *Proceedings of the 5th Symposium on Computer Arithmetic*, May 1981, pp. 81–86.

[41] Knuth, pp. 248–256.

[42] Pratt, V. R., *Course Notes for 6.046* (Cambridge, MA: MIT, 1975).

[43] Kornerup, P. and Matula, D. W., "An Integrated Rational Arithmetic Unit," *Proceedings of the 5th Symposium on Computer Arithmetic*, May 1981, pp. 233–240.

[44] Newman, W. M. and Sproull, R. F., *Principles of Interactive Computer Graphics*, Second Edition (McGraw-Hill, 1979), pp. 327–329.

[45] Robinson, J. B., *Modern Digital Troubleshooting* (Redmond, WA: Data I/O Corp., 1983).

[46] Siewiorek, and Swarz.

[47] Ledgard, H. F., *Programming Proverbs for FORTRAN Programmers* (Hayden, 1975).

[48] Polya, G., *How to Solve It*, Second Edition (Princeton University Press, 1973).

[49] von Oech, R., A Whack on the Side of the Head (Warner Books, 1983).

[50] de Bono, E., *The 5-Day Course in Thinking* (Penguin, 1967).

# 3
## Device-Dependent, Electrical, and Mechanical Rules

## 3.1 Device-Dependent Rules

This chapter looks closely at the different logic families and other devices that are our tools. These guidelines should simplify choices among different logic families.

**Logic Families.** Typically, a logic family is chosen before high-level design is begun. Often, it is wise to remember the existence of other families. The comparison in Table 3.1 may be helpful (note that numbers are only approximations).

A few of the terms in the table may be new. The "basic gate" is the fastest or simplest 2-input gate. Speed and power should both be minimized, but often one is traded for the other. The speed-power product is an indication of the overall "efficiency" of the family. A small number here says that the power consumed is put to good use. Wired functions are not available in all families but, when available, allow signals to be simply wired together to perform a logic function. Packing density is a measure of the number of gates that can be put on a piece of silicon of a given size. Complicated devices such as large dynamic RAMs and microprocessors are implemented in a family with a good packing density.

Let's summarize the characteristics of these families. TTL (Transistor-Transistor Logic) has many forms. It is the most popular logic family and has a wide range of applicability [3, 4, 5, 6, 8].

ECL (Emitter-Coupled Logic) is typically used in very high speed designs, where its high heat dissipation and power consumption can be tolerated [2, 3, 5, 6, 7].

CMOS (Complementary Metal-Oxide Semiconductor) is typically used where low power consumption is needed (e.g., battery-operated devices). CMOS is a combination of PMOS (Positive MOS) and NMOS (Negative MOS). New and popular high-speed CMOS families are also emerging, although Table 3.1 discusses only the slower 4000 family CMOS [1, 2, 3, 5, 6].

IIL (Integrated Injection Logic) is used in custom applications and is not available as a standard logic family. It has extremely low power dissipation and good packing density [2, 5, 6].

HTL (High-Threshold Logic) is used in high-noise environments [3, 6].

HiNIL (High Noise Immunity Logic), which is also called the 300 series, can switch unusually large amounts of current. It can drive 1000-foot lines. It is especially useful in noisy environments (due to the environment or inexpensive power regulation), such as those found in copy machines, vending machines, dictaphones, and other low-speed, high-noise devices [5, 53].

TTL is the most popular standard logic family and is expected to remain so through the 1980s. Of the types of TTL, low-power Schottky accounts for

| | TTL | ECL | CMOS | IIL | HTL | HiNIL |
|---|---|---|---|---|---|---|
| Basic gate | NAND | OR/NOR | NOR/ NAND | NOR | NAND | |
| Gate prop. delay (ns) | 3–30 | 0.75–2.0 | 30 + | 25–250 | 90 | 300 |
| Power dissip. per gate (mW) | 1–20 | 25–60 | 1 @ 1MHz, .0001 static | .001–.01 | 55 | 30 |
| Speed-power product (pJ) | 4–132 | 25–60 | 1 | 0.1 | 5000 | 9000 |
| Rise time (ns) | 1–5 | 0.6–2.0 | | | | |
| Input capacitance (pF) | 5 | 4–5 | 5 | | | |
| Fanout | 10 | 25–80 | 50 + | | 10 | 20 |
| Power supply (V) | + 5 | − 5.2 | 3–18 | | + 15 | 1–16 |
| Wired functions | AND (some) | OR | none | | | AND (some) |
| Noise generation | med-high | very low | low-med | low-med | med | low |
| Noise immunity | good | good | excellent | good | excellent | excellent |
| Cost | low-med | med | | | | |
| # devices available | very high | high | med-high | (custom) | med | low |
| Packing density | poor | poor | good | very good | | |
| Temperature stability | fair | poor-good | excellent | | | |

**Table 3.1** Comparison of logic families.

well over half of the TTL sales (measured in dollar volume). CMOS is the second most popular standard family, and its high-speed devices will make up more than half of CMOS sales in the latter 1980s. ECL is the third most popular family. The remaining families, HTL, HiNIL, and the older families such as RTL and DTL, together account for a tiny fraction of the market. Microprocessors and memories together, by the way, account for many times the volume of standard logic [40].

**Advanced Families.** Gallium arsenide (GaAs) has left the laboratory to become a commercially available fabrication process. It runs roughly an order of magnitude faster than ECL, has a speed-power product of 0.1 to 10 pJ, has a very wide operating temperature range, and is more radiation hard (immune to radiation) than silicon.

Josephson junctions are farther from the marketplace but promise greater speed than GaAs. This technology must be cryogenically cooled (that is, cooled to a temperature near absolute zero, $-273 \degree$C).

**TTL Comparison.** The 7400 TTL family was developed by Texas Instruments in the early 1960s. Two predecessor families, RTL (Resistor-Transistor Logic) and DTL (Diode-Transistor Logic) are obsolete. Table 3.2 compares the different types of standard 7400 TTL. The propagation delay is measured at 15 pF.

The 5400 TTL family contains the military versions of 7400 TTL parts. These military parts operate over a wider temperature range but may not meet the same performance specifications as their 7400 equivalents.

| | Avg. Propagation Delay (ns) | Power Consumption (mW) | Speed * Power (pJ) |
|---|---|---|---|
| Gold Doped | | | |
| 74 | 10 | 10 | 100 |
| 74L | 33 | 1 | 33 |
| 74H | 6 | 22 | 132 |
| Junction Isolated Schottky | | | |
| 74LS | 10 | 2 | 20 |
| 74S | 3 | 20 | 60 |
| Oxide Isolated Schottky | | | |
| 74ALS | 4 | 1 | 4 |
| 74AS | 1.5 | 10 | 15 |
| 74F | 3 | 4 | 12 |

**Table 3.2** Comparison of basic TTL families. The power consumption is per gate, and the speed-power product is a measure of the "efficiency" of the device [4].

The gold doped families are losing popularity to the newer Schottky families. Most, if not all, new logic devices are presented in these newer families.

**TTL Fanout.** Doublecheck the loading demands when driving a device from one TTL family with a device from another. Table 3.3 gives the worst case fanout capabilities of each TTL family driving each other family. "Worst case" here means that the smaller of the high level (logic 1) and low level fanouts was taken. The families along the left of Table 3.3 are the sources; those along the top are the families of the loads.

Note that no times are mentioned here. Fanout questions are twofold: 1) Can the output supply sufficient current for the loads? and 2) What does the line capacitance do to the rise/fall time? This table answers only the first question; that is, it reports how many loads *can* be driven, but it does not say *how long they will require.* One fact that can be added to Table 3.3 is that propagation delay depends in part on load capacitance (delay increases close to 1 nanosecond per picofarad of extra load capacitance). (Delay vs. capacitance is discussed in Propagation Delay Calculation in Section 2.4.)

**MOS Comparison.** Several different MOS variations exist — NMOS, PMOS, and CMOS (negative, positive, and complementary MOS) — and each can be fabricated with either metal or silicon gates. Combine this with the different manufacturing processes, and we have an extremely broad range of characteristics from which to choose. Various tradeoffs are made in power dissipation, packing density, speed, fanout, and so forth to arrive at MOS processes used in all types of memories, processors, and logic.

|        | **74** | **74L** | **74H** | **74LS** | **74S** | **74ALS** | **74AS** | **74F** |
|--------|--------|---------|---------|----------|---------|-----------|----------|---------|
| 74     | 10     | 80      | 8       | 40       | 8       | 40        | 4        | 10      |
| 74L    | 1.2    | 10      | 1       | 5        | 1       | 5         | 0.5      | 1.2     |
| 74H    | 12     | 50      | 10      | 25       | 10      | 25        | 2.5      | 12      |
| 74LS   | 5      | 40      | 4       | 20       | 4       | 20        | 2        | 5       |
| 74S    | 12     | 100     | 10      | 50       | 10      | 50        | 5        | 12      |
| 74ALS  | 2.5    | 5       | 2       | 10       | 2       | 20        | 2        | 2.5     |
| 74AS   | 12     | 111     | 10      | 50       | 10      | 100       | 10       | 12      |
| 74F    | 12     | 100     | 10      | 50       | 10      | 50        | 5        | 12      |

**Table 3.3** Fanout comparison of TTL families [14]. Watch out for delays beyond the times in the data book due to the capacitance of many inputs.

The largest use of MOS is in the manufacture of memories and processors. There are, however, two popular standard CMOS families, 4000 and 74HC. The 4000 family is described in Table 3.1 (see Logic Families, above).

The 74HC family is functionally compatible with the very popular 7400 TTL family and has capabilities similar to 74LS (low-power Schottky) TTL. There are several important differences between 74HC and 74LS, however. 74HC has twice the input capacitance (10 pF) and has superior noise immunity. While 74HC drives 74LS well, a translator is often suggested to assist 74LS driving 74HC. Another trait of 74HC is that its power dissipation is proportional to the operating frequency. While quiescent (idle) operation consumes only miniscule amounts of power, dissipation is proportional to operating frequency and beyond 3 to 10 MHz, 74HC holds no power advantage (the actual threshold depends on the device). Load capacitance can greatly affect power dissipation. For example, adding 30 pF to a 5 pF load typically doubles the power dissipation on the 74HC device driving the load. Finally, increased temperature increases 74HC propagation delays.

One final trait of CMOS is that of SCR (silicon-controlled rectifier) latch-up. Latch-up can destroy a CMOS device and happens when the IC's power or ground pins are outside their proper voltage ranges. The problem is a serious one. There are preventive techniques [17], and some fabrication processes are said to be immune to latch-up.

**ECL Comparison.** As seen in Table 3.4, 100K ECL is the highest performance ECL family and also solves the problems of voltage compensation and temperature compensation (that is, supply voltage noise and temperature minimally affect the signal levels and noise margins). It uses a smaller supply voltage for reduced power consumption but can interface with 10K or 10KH and can use a −5.2 V supply.

These ECL families use a standard package, where 7400 TTL and 4000 CMOS do not. There are exceptions to this rule, especially for memory, processors, and other devices not strictly in the 10K/100K numbering sequence, but most devices are housed in the same sized package.

**ECL Noise Margin vs. Temperature.** The 10K and 10KH ECL families lose significant noise margin when the temperature difference between a source and destination is greater than 35 degrees Centigrade [9]. 100K ECL has overcome this problem. Interfacing between 100K and either 10K or 10KH is best done with differential signals (signal pairs in which a voltage difference, not an absolute voltage, indicates the state of the signal) [22]. Differential signals are discussed in Inter-Board Communication in Section 3.3.

|                                      | **10K**       | **10KH**      | **100K**                        |
|--------------------------------------|---------------|---------------|---------------------------------|
| Avg. propagation delay (ns)          | 2.0           | 1.0           | 0.75                            |
| Per gate power Consumption (mW)       | 25            | 25            | 40                              |
| Speed * power (pJ)                    | 50            | 25            | 30                              |
| Rise Time (ns)                        | 2.0           | 1.5           | 0.8                             |
| Basic package                         | 16-pin DIP    | 16-pin DIP    | 24-pin flatpack or thin DIP     |
| Power supply (V)                      | −5.2          | −5.2          | −4.5                            |
| Voltage compensated?                  | No            | Yes           | Yes                             |
| Temperature compensated?              | No            | No            | Yes                             |

**Table 3.4** Comparison of ECL families.

**High-Level Family Differences.** Each standard logic family has a number of electrical characteristics that must be understood before design can begin. At a higher level, there are nonelectrical, logical characteristics that give a personality to each family:

- **TTL.** 7400 TTL has a very wide range of functions, speeds, and power consumptions to choose from. To widen the family further, some devices are ECL or unipolar internally but provide TTL-compatible signals. TTL uses whatever package size and shape best suits the task at hand. Open-collector outputs (for wire-ANDing signals) and tri-state outputs are available. Outputs that are neither open collector nor tri-state must not be tied together.

- **ECL.** 10K and 100K ECL must have its signals pulled down, which means that wire-ORing is possible on every signal. True and complementary signals are created equally quickly (not true with TTL), and complementary outputs are often provided. True and complementary outputs are required for differential transmission. The standard package size occasionally causes an unusual assignment of the signals; for example, the 100122 is 9 buffers (not 8, as one might expect) and the 10104 provides four 2-input ANDs, only one of which has a complementary output. An unusual device, the content-addressable memory (CAM) is supplied as device 10155.

- **CMOS.** In addition to the common devices found in the other families, 4000 family CMOS provides several shift registers with no parallel load and only the shift out value accessible (some have several signal taps at various stages). Wide counters with no parallel load are also provided, as is a Johnson counter.

The 74HC high-speed CMOS family has very similar characteristics to the 74LS (low-power Schottky) family. This family is logically identical to 7400 TTL.

**Tri-State Outputs.** All devices on a tri-state bus should be switched off *before* another bus is switched on. Unfortunately, strictly following this rule can be inconvenient in some applications. It seems that having two devices on together for perhaps five nanoseconds causes no harm, as long as the percentage of time that this happens is low. However, heat generation can greatly increase, and the large current drain may add noise to the power or ground planes (see Figure 3.1).

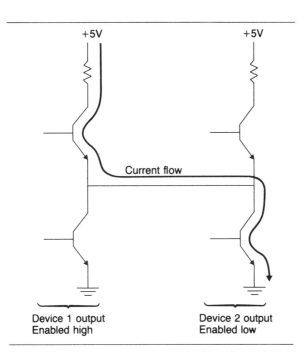

**Figure 3.1** Tri-State Outputs Proverb: There is a problem with having two tri-state TTL outputs concurrently enabled. As shown here, one output high connected to another that is low provides a fairly low impedance path from power to ground.

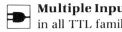 **Multiple Inputs.** Multiple inputs of a single device can be tied together in all TTL families except 74LS (it reduces the noisy immunity for 74LS) [11].

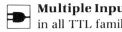 **Supply Noise.** ECL and TTL are more susceptible to noise in the ground input than in the power input [13].

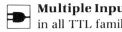 **Digital Trivia.** Have you ever wondered why some manufacturers use slightly different part numbers for equivalent devices? The "SN" from SN74S00 means "semiconductor network"; "DM" in DM74S00 stands for "digital monolithic" [51].

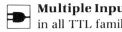 **Power/Ground Pins.** Examine the placement of power and ground pins in the pinouts of devices from unfamiliar families. For example, 10K ECL often has two ground pins and MOS dynamic RAMs have the +5 V and ground pins reversed from TTL's pinout (see Figure 3.2).

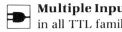 **Fast Clocks.** The rise time of 74AS (1 ns, typical) gives a faster edge rate than the rates of both 10K ECL and 100K ECL. Clocks in the fast TTL families over a certain length need to be terminated to minimize reflections. It is easy to think of terminations as being applicable only to ECL design and communications lines. However, do not overlook TTL designs for possible similar treatment. (See the more quantitative discussion in Impedance Mismatching in Section 3.3.)

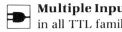 **Propagation Delay vs. Temperature.** The range at which most TTL propagation delays are fastest in room temperature (25°C). Note, however, that this is not always the case. The 74LS TTL high-to-low transition

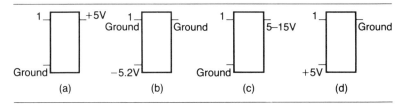

**Figure 3.2** Power/Ground Pins Proverb: Power/ground pinouts of several common devices: (a) 7400 TTL, (b) 10K ECL, (c) 4000 CMOS and (d) 64K MOS DRAMs.

gets *faster* as the temperature increases. ECL is also fastest at 25°C, but 4000 CMOS is fastest at the lower temperatures and becomes slower as the temperature rises.

**Static.** Some logic families are especially static-sensitive; these families include 100K ECL, MOS, and the high-speed TTL families such as 74F (FAST). Standard TTL and Schottky TTL are not immune to static damage, though they are much more rugged. When debugging a board, static-prevention techniques may not be used by designers as much as required. To compensate, the possibility of static damage should not be overlooked as an explanation to problems.

The following handling precautions reduces the number of device failures due to static:

- Always store chips in conductive foam or conductive plastic tubes.

- A technician handling chips should be grounded, as should the tools. The work surface should be conductive and should also be grounded.

- The work environment should be non-electrostatic. There should be no carpet and no electrostatic substances such as wool, silk, synthetic fabrics such as nylon, plastic tape, or styrofoam cups.

- Keep the humidity high in the work area (perhaps 65%) and consider the use of ionized air blowers for localized areas.

- Do not insert or remove devices or boards with the power on.

- All unused device inputs must be connected to power, ground, or the output of a logic device.

**TTL Inputs Tied High.** If a TTL input is to be tied high (pulled up), use the +5 V power supply through a resistor of 1000 Ohms or greater resistance. Do not tie TTL inputs directly to power since the chip power input can typically tolerate up to 7 V (which might occur on power up), while signal inputs must not exceed 5.5 V. The 74F family is an exception in that it can tolerate 7 V on both power and signal inputs. (For more information, see Pullups for Sets and Resets in Section 2.4 and Pullups in Section 3.2.)

**ECL Quieter.** Unlike TTL and CMOS, ECL current consumption does not change when switching. This eliminates sudden drains on the supply voltage and influxes of current into the ground. This is the reason for ECL's very low noise generation. By not making sudden demands for power as TTL does, ECL has reduced decoupling capacitor requirements.

**Mixing TTL Families.** Watch out for mixing TTL families. This is especially true for 74LS with the faster families (74S, 74F, and 74AS). The problems arise from impedance mismatches between the families. If possible, use isolated ground and power planes for 74LS (see Figure 3.3) [11].

**Separate Grounds.** Because ECL perturbs the ground less than TTL and has a smaller tolerance of noise, try to separate the grounds used by sections of ECL and TTL on the same board. Of course the grounds are connected, but they are attached only at small points, hopefully near capacitors. This technique can also be used to isolate the ground of a 74LS section from higher-speed TTL (see Figure 3.3).

**RAS Signal for DRAMs.** The input signals to dynamic RAMs, such as the Row Address Strobe ($\overline{RAS}$), must meet the timing requirements at all times. A truncated $\overline{RAS}$ might mean more than simply that the access being performed will not be complete — it might cause damage to data in the RAM. For this reason, synchronize the resetting of the RAM signal generator to the RAM access cycle so that no truncated cycles occur. By the way, negative logic signals are often documented with a slash at the end of their names ($\overline{RAS}$ = RAS/).

**Dynamic RAM Initialization.** Those few applications that allow the dynamic RAM (DRAM) chip refreshing to lapse may have to reinitialize the DRAMs with 8 $\overline{RAS}$ cycles *before they can be used* [12, 15].

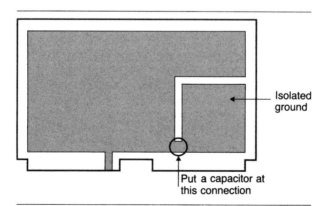

**Figure 3.3** Mixing TTL Families Proverb: Isolated ground plane used for protecting 74LS (or ECL) from the noise generated by the faster TTL families.

**IC Derating.** "Derating" means reducing the stresses placed on a device below the limits the manufacturer guarantees. While linear devices can be derated to increase life (for example, reducing the voltage across a capacitor or the wattage dissipated by a resistor), digital integrated circuits typically can not tolerate a voltage reduction and do not need reductions in AC characteristics (rise time, propagation delay, and so forth). However, the receivers and drivers on chips can degrade with time, and a 20% reduction in fanout and a 10% reduction in noise-margin levels are recommended [16]. Without this derating, old age failure may be a more serious problem (see the discussion of the bathtub curve in Error Sources in Section 2.3).

**Device Failure Rates.** Two IC failure rates are important. The $\lambda$ of a device records the rate of failure and is usually measured in failures per million hours of operation. This is the "reliability" of the device. Another way of looking at the same information is to examine the mean time between failures (MTBF), which is the average time from one failure to the next. The MTBF is computed $1/\lambda$.

There are two ways of calculating $\lambda$. MIL-HDBK-217 is a military failure rate prediction guideline [54]. It uses a base failure rate and also considers the additional stresses of shock, humidity, vibration, temperature, and lack of maintenance, which are found in various environments. A second method of determining the $\lambda$ is through parts testing, such as performed and reported by the Reliability Analysis Center of the Rome Air Development Center.

The second failure rate is the rate of failure of devices arriving from the manufacturer, usually measured as a percentage. This percentage is often called an Acceptance Quality Level (AQL) or, simply the "quality." For example, an AQL of 1% means that 10 devices out of every 1000 will, on the average, not meet all the specifications to which they are guaranteed. An AQL of 0.1% for all types of devices (memory, logic and processors) is today considered excellent.

Product quality is becoming so high that a major cause — if not *the* major cause — of device failure is improper handling of parts after they leave the manufacturer, either by a distributor or by the end user (see also Date Codes in Section 5.3).

**Composite Failure Rate.** A chip failure rate from the manufacturer even as high as 1% means that "few" of any collection of chips are bad; more valuable would be some idea of how the failure rate of individual chips combine to produce the composite failure rate of a board. This composite failure probability is $1 - (1 - p)^n$, where $p$ = probability of failure (also referred to as Acceptance Quality Level [AQL]) and $n$ = number of devices. For example, a failure rate of 1% for each chip on a 200-chip board gives $1 - (1 - .01)^{200} = 0.87$, or 87% as the likelihood of one or more bad chips on the board. If the probability of failure is 0.87, the probability of success is $1 - 0.87 = 0.13$. A 10-fold improvement in the failure rate gives $1 - (1 - .001)^{200} = 18\%$ (i.e., fewer than 1 board in 5 will have a chip-related error). The cost of finding a bad chip by debugging underscores the importance of eliminating as many chip failures as possible before board manufacture (and preventing more failures during manufacture) [18, 52]. The composite failure rate of many chips with different failure rates is computed: $1 - [(1 - p_1)^{n_1}(1 - p_2)^{n_2}...(1 - p_i)^{n_i}]$.

**Junction Temperature.** Digital devices fail if stressed with too high a temperature. However, although the case temperature (external temperature) of the device is easy to monitor, it is the internal junction temperature of the device which must be kept below a maximum. (Junctions are the interfaces which form the transistors.) The junction temperature is calculated:

$$T_J - T_a + p * \Theta$$

where $T_J$ = junction temperature (degrees C)
   $T_a$ = ambient temperature (air temperature in degrees C)
   $p$ = power dissipated by the device (W)
   $\Theta$ = thermal resistance (degrees C/W).

The junction temperature must be kept below the maximum defined by the manufacturer. The thermal resistance $\Theta$ can be thought of as the heat insulation characteristic of the device's case — the poorer an insulator and the lower the thermal resistance, the lower the junction temperature. The thermal resistance is lowered by using a ceramic instead of plastic case, by using a case with a larger area, by using heat sinks to increase the area even further, and by using fans to move the air around the device. This resistance can range from $200°/W$ or more for a small plastic case to $50°/W$ or less for a large ceramic case with heat sink [45]. (For more information, see Fans in Section 3.2.)

## 3.2 Electrical and Mechanical Considerations

These are some of the low-level mechanical and electrical concerns.

**Care of Capacitors.** Because the failure of capacitors is difficult to measure but can seriously increase power and ground noise, it is important to take care of them. A voltage derating of up to 50% is recommended [19]. This means that the peak voltage transient expected (perhaps 7 volts with a 5-volt power supply) should be 50% of the capacitor's breakdown voltage. This suggests a 15 V or greater breakdown voltage for TTL and ECL capacitors, both bulk and decoupling (see Decoupling Capacitors in Section 3.3.)

Heat is also a major cause of capacitor failure. A general guideline states that each 10°C increase in operating temperature reduces the capacitor's life expectancy by 50% [20].

**Pullups.** Pullup resistors can be used to maintain a signal permanently high. They might be used, for example, to disable the SET or RESET inputs to a flip flop. The following example shows how to determine the maximum number of signals that can be tied to a single pullup resistor. The example assumes that the pullup is used to disable the RESET ("CLEAR") inputs of 74S74 flip flops (see Figure 3.4):

1. A Schottky TTL data book reports that the high-level input current of the 74S74 RESET input (the current drawn when a high level is presented) is 0.15 mA.

2. Anticipating worst-case conditions, we assume that the voltage supply is at its minimum, 4.75 V. Our goal is to maintain the voltage at the input above the minimum high-level voltage which, for this device, is 2.0 V. To simplify the calculation, however, we use 2.7 V, the voltage at which the high-level input current is measured. The maximum voltage drop allowable is then $4.75 - 2.70 = 2.05$ V.

3. Ohm's law states that the current flowing through a resistor is computed $I = V/R$. Letting $R = 2K\,\Omega$, for example, $I = 2.05/2000 = 1.025$ mA. This is the maximum current we can allow to flow through the resistor; any more might bring the voltage at the inputs too low.

4. With each RESET input drawing 0.15 mA, there is enough current for $1.025/0.15 = 6$ inputs.

5. The heat dissipated by the resistor is computed: $W = VI = 2.05\text{ V} \times 0.001\text{ A} = 2$ MW. This is the minimum wattage rating of the resistor.

6. Halving the resistance to $1K\,\Omega$ doubles the number of RESET inputs allowable on this pullup to 13 and similarly doubles the maximum heat dissipation.

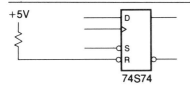

**Figure 3.4** Pullup Proverb: Pullups of 74S74 flip flop RESET input.

Setting ECL inputs to a constant high or low is easier since, when left unconnected, they float to a solid low and can be pulled up by a connection to ground. (For more information, see Pullups for Sets and Resets in Section 2.4, TTL Inputs Tied High in Section 3.1, and Pullups and Grounds in Section 6.2.)

**Pulled-Up Lines on a Bus.** Where a number of boards put signals onto an open-collector (TTL) or open-emitter (ECL) bus, the pullup/termination is best done 1) once on the bus, 2) at one source, or 3) at one destination (the third choice provides the best termination). Watch out for accidentally terminating a signal more than once. A multiply-terminated signal is effectively terminated with a lower resistance than was intended (terminating a line m times with n $\Omega$ each time is equivalent to terminating it once with a resistance of n/m $\Omega$). A lower termination resistance means that the line drivers will run hotter and the line may not be terminated at its characteristic impedance.

**Radiation Emission.** New Federal Communications Commission (FCC) rules (docket 20780, part 15) set strict radiation emission limits on Class A (industrial) and Class B (consumer) devices using clocks over 10 kHz. This radiation is called Electromagnetic Interference/Radio-Frequency Interference (EMI/RFI). Class B limits are more strict. In Europe, other standards-setting organizations have defined similar limits on EMI/RFI.

The military's TEMPEST requirements (defined in NASCEM 5100), are imposed on some systems processing classified information. The elimination of intelligible electronic noise by TEMPEST shielding prevents eavesdropping.

The easiest technique to reduce radiation emission after the system has been designed is to improve the chassis shielding. Aluminum protects against electric fields, but only ferrous metals also protect against magnetic fields [21]. Power filters and cable shielding are also important. FCC compliance is usually a concern only during system design and integration. Shielding need not affect the design of a board, but proper use of capacitors, the minimization of the clock frequency, and the use of ferrite beads can be valuable in reducing radiation at its source. Ferrite beads are small, inexpensive sleeves that fit over conductors. They sharply suppress the frequency content over roughly 1 MHz in the conductor. (For more information, see Drivers/Receivers in Section 3.4.)

**Ground Plane.** It is best to put the ground plane on the component side of the board [23]. Be sure that a wire-wrap board is fabricated with an insulator (usually a thin plastic coating) over both the power and ground planes. It is not wise to have either plane exposed to oscilloscope probes, oscilloscope grounds, and stray wires.

**Heat Sinks.** ECL and Schottky TTL are candidates for heat sinks. For example, a 74S299 8-bit TTL shift register is only a 20-pin device but has a maximum heat dissipation of 1.2 watts. Make sure that the orientation of the heat sink puts the fins parallel to the air flow. One source recommends heat sinks for devices dissipating more than 750 mW [24]. Without a heat sink, hot chips may suffer from reduced reliability. (For more information, see Junction Temperature in Section 3.1.)

**Sockets.** Use sockets in production boards (boards being produced in quantity) appropriately. PROMs, failure-prone devices, non-wave-solderable devices, and devices that may later be replaced with a pin-compatible upgrade should be in sockets. Microprocessors can be put in sockets to allow the use of emulators or other software development tools. However, do not use sockets unnecessarily. Sockets increase board cost, increase labor costs, increase signal capacitance and inductance by increasing the lead length, can drastically reduce reliability, and raise the profile of its device above that of neighboring devices. Not having uniform profiles distorts air flow, possibly causing hot spots, and leaves the tall devices more vulnerable to abuse. (For more information, see Device Profiles in Section 4.2.)

**Fans.** In a chassis with circular fans, different card slots often have different cooling characteristics. A fan best cools the board nearest its center. The hotter boards should be placed in the coolest card slots. A chassis modification that reduces the problem of inferior (hotter) card slots puts a several-inch-wide plenum chamber between the output of the fans and the card assembly. This gap distributes the air flow more evenly. The fans are typically the only part of a system which make audible noise. Investigate the intended application environment to find the maximum allowable noise. (For more information, see Ergonomics in Section 2.1.)

Special cooling problems may require nontraditional tools. Piezoelectric fans are very small when compared with traditional circular fans and can be placed near a problem area. Vortex tubes use a source of compressed air to supply chilled air.

Ideally, the system should be designed to run without the fans, so that the failure of the fans does not lead to the failure of the system. In this case, the fan's purpose is to increase reliability by lowering the temperatures of the electrical devices — the rates of many types of device failure are proportional to the device's temperature. (For more information, see Junction Temperature in Section 3.1.)

## 3.3 Wiring and Noise

The analog nature of digital design may be most visible in the area of signal transmission. These suggestions deal with transmission problems and their solutions.

**Frequency Content.** A data stream is labeled a "1 MHz" stream if it will potentially change state once every microsecond. Note, however, that a 1 MHz clock has a rising edge once every microsecond but changes state *twice* every microsecond. Because clocks change twice as fast as their associated data streams, they can be the signals that suffer the most during transmission.

Frequency content in the transmission of digital signals is more often a function of the rise/fall time of the signal transitions than the frequency of the data in the signal. The signal's frequency understates the actual frequency composition of the transmitted signal because the signal is not a sinusoid, but an approximation to a square wave. Square waves, even imperfect ones such as digital clocks, have frequency components are greater than the "clock frequency." The signal's rise time is related to its frequency only in that the frequency sets a lower bounds on the rise time.

**Impedance Mismatching.** TTL and ECL devices have low output impedances and high input impedances. This combination can cause ringing in signals. Instead of terminating lines to eliminate reflections (which is usually done with ECL), the line length can be shortened so that the one-way propagation delay is less than $1/4$ the rise/fall time of the signal [25]. The line lengths implied by this rule are shown in Table 3.5 (assuming 1.5 ns propagation delay per foot of wire) [9, 27].

Clock lines violating these rules are candidates for termination. Data lines are more immune to this problem if there is sufficient time for the reflections caused by the impedance mismatch to die out before sampling at the destination. Suprisingly, as Table 3.5 shows, 74S and the faster TTL families have faster edge rates the 10K ECL families, since the ECL voltage swing is less than half that of TTL. (For more information, see Fast Clocks in Section 3.1.)

| Logic Family | Typical Rise Time (ns) | Maximum Unterminated Wire Length (inches) |
|---|---|---|
| 100K ECL | 0.8 | 1.6 |
| 74AS | 1.0 | 2 |
| 74F (FAST) | 1.5 | 3 |
| 74S | 1.5 | 3 |
| 10KH ECL | 1.5 | 3 |
| 10K ECL | 2.0 | 4 |
| 74LS | 5.0+ | 10 |
| 4000 CMOS | 20.0+ | 40 |

**Table 3.5**  Maximum unterminated wire length for each logic family.

**Signal Interconnection Method.** A daisy-chain interconnection from the source to the destinations is preferred to a tap-off or star network (see Figure 3.5). If stubs are necessary (that is, if T's in the wire are necessary), control the stub length such that the propagation delay of the stub is less than $1/8$ of the signal rise time on the line [29]. With ECL and the faster TTL families, this rule limits stubs to less than two inches.

**Transmission Resonance.** "The maximum length of a wire which can effectively equalize voltage differences between its extremities is $1/10$ to $1/20$ of a wavelength" of the signal travelling on it [30]. When the wire length is near one of certain multiples of the wavelength, such as $1/4$, $3/4$, $5/4$, and so forth, the signal reflections can cause resonance and cause the wire to become an ineffective conductor. A resonant conductor can radiate energy and become a

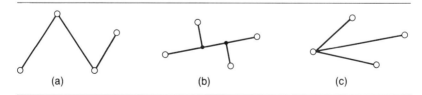

(a)        (b)        (c)

**Figure 3.5**  Signal Interconnection Method Proverb: Signal interconnection methods (a) daisy chain, (b) tap-off, and (c) star. Note the stubs in the latter two methods.

noise source, itself. Additionally, a conductor can serve as an antenna for certain frequencies. Line termination and shielding limit these effects.

By the way, the wavelength of 2 MHz, 10 MHz, and 50 MHz signals are approximately 750, 150, and 30 feet long, respectively.

 **Line Termination.** Termination prevents signal reflections on a line. Several types of signal termination are available (see Figure 3.6) [25, 31].

- **Series.** This termination can drive a star network well and discourages crosstalk better than parallel termination. It works by attenuating higher frequencies.

- **Parallel.** The propagation delays of drivers and the edge speeds are not affected, but the propagation delay of the line is increased.

- **Diode.** This type of termination is easy to implement because the impedance of the line does not affect the choice of diodes. It is valuable for clipping overshoot and undershoot (some devices already have input diodes for clipping undershoot). Schottky diodes are usually used.

**Inter-Board Communication.** When sending signals between boards, consider the techniques in Table 3.6 when choosing the proper tradeoff between economy, reliability, and ease of use.

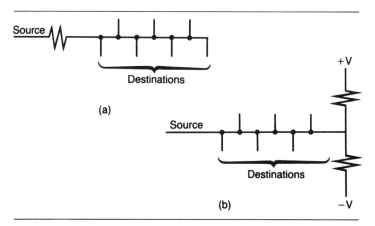

**Figure 3.6** Line Termination Proverb: Line termination techniques: (a) series and (b) parallel.

| Technique | Comments |
|---|---|
| Individual Signals | |
| printed circuit (PC) | Simple; requires no installation. |
| PC over ground plane | Less noise than PC but more expensive. |
| wire wrap | Some effort to install but interconnections can be customized and updated easily. |
| grounded twisted pair wire wrap | A ground wire is twisted around the signal wire. Less noise than wire wrap. |
| differential wire wrap | A signal is converted into two signals: an equivalent of the original and its complement. A twisted pair is used to send the signals. More noise immune than twisted pair because only the *difference* between the signals, not their *values*, is important. [32]. |
| coaxial cable | A ground shield encloses the signal. Very good noise immunity. |
| fiber optics | Absolute noise immunity. |
| Ribbon Cable | |
| plain | Simple, economical. |
| with ground plane | A ground shield protects the wires, similar to the PC with a ground plane. |
| with ground signals | Alternating ground signals reduce crosstalk. Only half the wires are available for signal transmission. |
| twisted pair | Pairs of wires are twisted, allowing them to be used as twisted pairs (with one wire ground) or differential pairs. |

**Table 3.6** Communication techniques.

Other techniques for reducing noise are hysteresis on receivers and increasing driver output voltage to drown the noise by increasing the signal-to-noise ratio [33]. (For more information, see Bends in Signals, Keyed Connectors, and Crosstalk in Section 4.2.)

**Grounding for Transmission.** When driving a signal with an attached ground (such as twisted pair or ribbon cable with ground), it is recommended that the driver/receiver device ground, the decoupling capacitor ground, and the transmission-line ground be brought to a common point [34]. This ensures that all "grounds" are as equivalent as possible.

**Fiber Optics.** Fiber optics have a number of advantages over coaxial cable and other electrical low-noise transmission techniques [37]:

- High bandwidth with little attenuation. Where coaxial cable is typically limited to several hundred feet for 40 MHz signals, fiber optics can send 60 MHz signals over 5000 feet reliably and without repeaters.

- No noise susceptibility.

- No hum (60 Hz noise) and no ground loops.

- No radiation of noise, which is good both for FCC compliance and for TEMPEST (military) applications.

- Electrical isolation between source and destination. Damaging power or ground spikes, such as those caused by lightning, are prevented from propagating. The source and destination can have different ground potentials.

- The cable is smaller and more mechanically flexible than wire cable.

The disadvantages are:

- More expensive.

- Few devices are available, compared to digital drivers and receivers.

- More difficult interconnection techniques. While this has been improved, interconnection remains time consuming.

- Less widely used. There is a smaller base of experience from which to draw.

- There are limitations on the radius of curvature (bendability) of the cable.

- A nuclear electromagnetic pulse (EMP) can damage the fiber optics.

The difficulties with fiber optics will lessen over time. As a result, expect fiber optics to gain in popularity.

**Signal Capacitance.** Wires on a Multiwire board (Multiwire is a board fabrication technique) have a capacitance of 36 pF/foot with a 50 Ω impedance for the lowest layer and a capacitance of 27 pF/foot with 75 Ω impedance for the second layer [38]. Crossover wires add up to 20% to the capacitance. Note that the 50 Ω and the 36 pF/foot values for the lowest layer assume a *solid* ground/voltage plane; less copper means that the capacitance drops and the impedance rises. Capacitance on printed circuit traces is approximately the same as that for the second layer of Multiwire [39] (see Board Manufacturing Techniques in Section 4.1).

Do not overlook this signal capacitance: many device output signals are rated at only 15 pF. For example, a signal rated at 15 pF could not be expected to meet its rated propagation delay, if it were driving a single load four inches away.

**Decoupling Capacitors.** Digital designs need capacitors to stabilize the power plane. These capacitors are analogous to shock absorbers, damping out the perturbations in the power and ground planes caused by chips' varying power demands. Different-sized capacitors are used to dampen the different frequency components of noise. Table 3.7 is a rough guide to the requirements of several logic families for small, high-frequency capacitors.

Several references discuss the requirements for decoupling capacitors for Schottky TTL [28, 41, 42], the newer TTL families [8, 43], ECL [35, 44], MOS dynamic RAMs [49], and 4000 CMOS [36]. LSI and VLSI devices need decoupling also; one capacitor per power pin is a good rule to follow.

These decoupling capacitors are physically quite small (some are packaged as 2-pin DIPs). They must be able to respond to very high frequency perturbations on the voltage and ground planes and so are usually high-quality, low-inductance ceramic capacitors. Their values usually lie between 0.01 to 0.1 microfarads. Balph [46] suggests mixing the values within this range throughout the board. With a poorly laid-out power or ground plane, more capacitors should be used (with one capacitor per chip, each chip has a dedicated "power supply"). The placement should be as near as possible to the chip. The electrical path from the capacitor, through the power plane, and into the IC's power input is typically responsible for twice the inductance of the

| Family | Number of Chips per Capacitor |
|---|---|
| 74LS TTL | 1 for drivers/receivers, 5 for others |
| 74S TTL | 1 for buffers, 2 for other devices |
| 74F TTL | 1 for buffers, 2 for other devices |
| 74ALS TTL | 2 to 5 |
| 74AS TTL | 1 for buffers, 2 for other devices |
| 10K ECL | 4 |
| 10KH ECL | 1 to 4 |
| 100K ECL | 1 to 2 |
| MOS dynamic RAMs | 1 |
| 4000 CMOS | 1 for buffers, 10 for other devices |

**Table 3.7** Thermal Relief Proverb: Decoupling capacitor requirements for common logic families. The suggested capacitance value ranges from 0.01 to 0.1 microfarads (except that the recommended value for 4000 CMOS is 3 microfarads).

capacitor itself. Close placement helps minimize this inductance and maximize the benefit of the capacitor [47].

Line drivers and receivers often need a dedicated decoupling capacitor. Synchronous TTL devices (especially counters and shift registers) are potential noise generators and might also need a dedicated capacitor [41].

Bulk capacitors (10 – 100 microfarads in size and usually made of tantalum or aluminum) should be used on the board at the input of each power supply. This minimizes noise entering or leaving the board through the supply inputs. Additionally, blocks of dynamic RAMs require one of these for every eight devices [48, 49]. Like the decoupling capacitors, these must be mixed throughout a dynamic RAM array. Putting the bulk capacitors together in one or two neat rows or columns sharply reduces their benefit.

When in doubt, put capacitor holes in a spare location along with the device holes. This space can then be used for a capacitor, if needed. Be conservative: anticipate using plenty of capacitors. Excess capacitors can always be deleted later to reduce the board cost.

The devices used have an effect on the noisiness of the power and ground planes, of course — the more sudden their surges from the power plane or to the ground, the worse the problem (see ECL Quieter, Tri-State Outputs, and Logic Families in Section 3.1). In addition to decoupling capacitors, there are some architectural considerations that can help this problem (see Sequential vs. Parallel in Section 2.2).

# 3.4 Special Devices

Several unusual devices are noted here (be sure to add more valuable devices to your notebook as you find them).

**Capacitors.** Consider Micro-Q capacitors from Rogers Corp. They fit under the chip, taking up little space and providing very little inductance (less than one tenth that of standard methods). They better quiet both the power and ground inputs to the chip better than traditional capacitors [50, 26]. Usually, these capacitors are more difficult to install than conventional capacitors (especially in sockets), since both the chip and the capacitor are installed together.

**Drivers/Receivers.** Consider the National DS3662 trapezoidal bus driver (it can handle up to a 20 MHz clock). This device uses special techniques to provide high noise immunity. By slowing the rise and fall times of signals, this device also reduces radiation emission. Using a regular gate or flip-flop output as a driver instead of the DS3662 or a line driver like a 75138 may compromise the quality of the signal [10].

**Register with Clock Enable.** Some members of Advanced Micro Devices' 29800 family of registers have a clock enable, which is valuable for cleanly qualifying the master clock.

# References

[ 1] *Understanding CMOS* (Somerville, NJ: RCA Corp., 1974).

[ 2] Fink, D. G. and Christiansen, D., *Electronics Engineers' Handbook*, Second Edition (McGraw-Hill, 1982), pp. 16–19 and 3–49.

[ 3] Milnes, A. G., *Semiconductor Devices and Integrated Circuits* (VNR, 1980), pp. 479 and 500.

[ 4] *Semiconductor Product News* (Dallas: Texas Instruments Inc., December 15, 1982).

[ 5] Mazda, F. F., *Integrated Circuits* (Cambridge University Press, 1978), pp. 32–33.

[ 6] Fink and Christiansen, pp. 8–80 – 8–89.

[ 7] Blood, W. R. et al., *MECL System Design Handbook*, Revised Edition (Phoenix, AZ: Motorola Semiconductor Products Inc., 1980).

[ 8] *FAST Circuits and Applications Seminar* (Mountain View, CA: Fairchild Corp., 1982).

[ 9] Balph, T., "Implementing High-Speed Logic on Printed Circuit Boards," *Proceedings of Wescon 81*, September 1981, p. 1.

[10] Balakrishnan, R. V., "Cut Bus Reflection, Crosstalk with a Trapezoidal Transceiver," *EDN*, August 4, 1983, pp. 151–156.

[11] *TTL Data Manual 1984*, (Sunnyvale: Signetics Corp., 1984), p. 3–6.

[12] "Intel 2164A 64K Dynamic RAM Device Description," *Memory Components Handbook* (Santa Clara, CA: Intel Corp., 1984), p. 3–30.

[13] Blood et al., pp. 193–194.

[14] *Semiconductor Product News* (Dallas: Texas Instruments Inc., January 1982).

[15] Reddy, A. et al., *64K Dynamic RAM Component Design Manual* (Phoenix, AZ: Motorola Inc., 1981), p. 8.

[16] Fink and Christiansen, p. 28–27.

[17] Mannone, P., "Careful Design Methods Prevent CMOS Latch-up," *EDN*, January 26, 1984, pp. 137–152.

[18] Blakeslee, T. R., *Digital Design with Standard MSI and LSI* (John Wiley & Sons, 1979), p. 355.

[19] Fink and Christiansen, pp. 28–20 – 28–21.

[20] Bennett, E., "Determining Switching Power Supply Reliability," *Electronic Products*, September 30, 1983, pp. 77–81.

[21] Williams, M. and Miller, S., *Series 54ALS/74ALS Schottky TTL Applications* (Dallas: Texas Instruments Inc., # B215) p. 20.

[22] *F100K ECL User's Handbook* (Mountain View, CA: Fairchild Corp., 1982), p. 3–20.

[23] Blood et al., p. 186.

[24] Ibid., p. 113.

[25] True, K. M., "Transmission Line Interface Elements," *The TTL Applications Handbook* (Mountain View, CA: Fairchild Corp., 1973), Chapter 14.

[26] *Q/Pac User's Guide* (Mesa, AZ: Rogers Corp.).

[27] True, p. 15–9.

[28] *Bipolar Microprocessor Logic and Interface Data Book* (Sunnyvale: Advanced Micro Devices Inc., 1985, pp. 8–13.

[29] True, p. 14–5.

[30] Kalbach, J. F., "Designer's Guide to Noise Suppression," *Digital Design*, January 1982, pp. 26–35.

[31] Blood et al., pp. 53 and 71.

[32] Ibid., p. 85.

[33] True, p. 14–4.

[34] Williams and Miller, p. 24.

[35] *F100K ECL User's Handbook* (Mountain View, CA: Fairchild Corp., 1982), p. 6–9.

[36] *Philips LOCMOS HE4000B IC Family* (Sunnyvale: Signetics Corp., September 1981), p. 90.

[37] "Fiber Optics in RGB Color Computer Graphics Communications" (Worcester, MA: Artel Communications Corp.).

[38] "High Frequency Propagation Characteristics of Multiwire Circuit Boards," (Glen Cove, NY: Multiwire Div. 1972).

[39] Peatman, J. B., *Digital Hardware Design* (McGraw-Hill, 1980), p. 51.

[40] "U.S. Semiconductor Consumption," *Electronics*, January 12, 1984, pp. 138–139.

[41] *TTL Data Manual 1984* (Signetics), p. 3–5.

[42] True, p. 15–11.

[43] Williams and Miller, p. 17.

[44] Blood et al., p. 25.

[45] *Digital IC Handbook* (Irvine, CA: Plessey Semiconductors, May 1982), pp. 46–49.

[46] Balph, p. 2.

[47] Martin, A. G., *Decoupling: 16K and 64K Dynamic RAMs* (Myrtle Beach, SC: AVX Corp.), p. 11.

[48] *Memory Handbook 1983*, Second Edition (Santa Clara, CA: OKI Semiconductor, 1983), p. 247.

[49] Reddy et al., pp. 24–25.

[50] *Memory Handbook 1983* (OKI) p. 252.

[51] Taub, H. *Digital Circuits and Microprocessors* (McGraw-Hill, 1982), p. 85.

[52] Siewiorek, D. P. and Swarz, R. S., *The Theory and Practice of Reliable System Design* (Digital Press, 1982), p. 11.

[53] *Data and Design Manual* (Mountain View, CA: Teledyne Semiconductor, May 1981), chapter 12.

[54] Siewiorek and Swarz, pp. 709–719.

# 4
# Manufacturing and Financial Considerations

Many designers are not accustomed to thinking of manufacturing and financial considerations as their concerns. The smallest organizations have so few people that the engineer must be a jack-of-all trades, but it is assumed that larger organizations are well-staffed for each task. In fact, regardless of the size of the work environment, many disciplines overlap with design, and the designer must know them to produce a product that does not fall short after it leaves the lab.

## 4.1 High-Level Philosophy

What needs to be done to reliably and repeatably manufacture a circuit board? Let's look at some high-level guidelines. (For more information, see Service Frequency in Section 2.3.)

**Board Manufacturing Techniques.** A number of techniques exist for the manufacture of circuit boards and for the laying of wires used to interconnect devices.

- **Printed Circuit (PC).** This is the least expensive technique. The PC board has thin copper paths (traces) that interconnect device leads and provide power and ground connections. Masks are clear plastic sheets with opaque ink or tape defining traces. Photographic techniques and the masks are used to deposit the copper, which forms the traces. To maximize the number of traces put on a single board (and thus maximize the number of possible chips), the horizontal traces are placed on one side, the vertical traces are on the other, and copper-plated holes interconnect signals on the top and bottom layers.

- **Multilayer.** An extension of PC, multilayer uses more than just two layers of signals. Layers are added as the chip density increases.

- **Multiwire.** Multiwire is an excellent technique when the chip density is high (when the inter-chip spacing is roughly 0.2 inches or less). Signal paths are thin wires that are laid into an epoxy base. A two-layer Multiwire board might hold as many paths as a 10-layer multilayer board. Multiwire can be faster into production than multilayer and is most economical for small-to-medium volumes of a high-density board.

  Multiwire is inherently a computer-automated process. Changes are simple, quick, and reliable because only the wire list is updated. New masks reflecting the changes must be made for new PC or multilayer boards. The PC and multilayer processes can be partially or fully automated, but the labor required in making manual changes is costly and error-prone.

- **Wire Wrap.** Wire wrap is a very convenient interconnection technique, which is often used to develop prototypes. A PC board supplies power and ground, and the long posts of the sockets holding the chips can be interconnected by wires. Since the wires can be laid point-to-point instead of only vertically or horizontally, wires tend not to run parallel for long distances, which minimizes crosstalk as well as their length. On the other hand, the wires typically have considerable slack, which increases the length. Since these wires are more flexible than copper traces, wire wrap can be a more rugged fabrication method than PC or multilayer.

While wire wrap is relatively immune from crosstalk, it is the worst technique in terms of susceptibility to other types of signal noise. The solid power and ground layers of Multiwire and, often, multilayer quiet most of this. Unfortunately, the solid power and ground also increases signal capacitance (see Signal Capacitance in Section 3.3).

**Chip Location Labelling.** There are two basic ways of defining chip locations on a board. The first typically assigns a unique number/letter combination to each chip location, while the second defines a grid by rows labelled with letters and columns labelled with numbers. The second way has the advantage that finding chips is easy, and only the board edges must be reserved for labelling (see Figure 4.1). On the other hand, this grid works best for components that fit well into a regular, rectangular layout (see Silkscreen in Section 4.2 and Row Letters in Section 5.3).

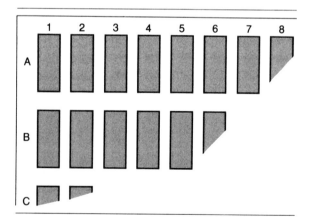

**Figure 4.1** Chip Location Labelling Proverb: Section of board showing rows and columns labelled with the silkscreen.

**Form Factor.** When deciding on a new custom form factor (board size and shape), consult the baseboard manufacturer to find out how efficient a proposed shape is to manufacture. Some shapes can be more expensive than would be expected from their size since they are wasteful in manufacturing.

A smaller board is more rugged. A stiffener across the unsupported edge can also increase board strength and reduce warping. Also, consider the longer board edge for the backplane connection. This increases board I/O potential (see Device Profiles in Section 4.2).

**Cost Estimate.** The designer should know the ballpark cost of all devices used. An estimate of the total parts cost (including the baseboard, integrated circuits, capacitors, and linear components) as well as unusual manufacturing requirements should be made when possible. This way, unpleasant surprises are avoided at manufacturing time.

A quick and easy rough guess of the sale price of the entire board is made by multiplying the number of integrated circuit packages expected by a dollar factor (typically $10 – $20). Donn [7] suggests $16 per IC. This factor can be modified knowing the dollars/IC ratios of other similar products. The factor drops when high volumes are expected and rises when low volumes or extensive research are likely.

**Minimize Chip Count.** This rule has its exceptions: sometimes, it is best to replace one expensive chip with two inexpensive chips. However, minimizing chip count usually tends to minimize the entire project costs. Chip count is roughly proportional to the following costs, all of which should be minimized:

- **Continuous factors:** IC ordering and inventory, IC testing, manufacturing time, decoupling capacitor purchasing and installation, documentation, and maintainance. A single extra chip slightly increases these costs.

- **Discrete factors:** board area, board testing, chassis size, power supply size, cooling requirements, and system integration and testing. Usually, a single extra chip squeezed onto an existing board, does not force the use of a larger power supply, and so forth. These costs do not increase continuously but in jumps. However, they are attributable to the chip count.

These costs imply an overhead cost per IC, varying according to the application but typically adding up to several dollars. In deciding whether to replace two $0.20 SSI devices with one $1.50 MSI device, don't forget the overhead. Guessing a $2.00 overhead cost, the SSI solution costs 2($0.20 + $2.00) = $4.40, while the MSI solution costs $1.50 + $2.00 = $3.50 [1, 4, 7].

Another benefit to a reduced chip count is that design and debugging time usually decrease. This benefit is real, although often not as great as might be expected. Do not forget that the more LSI and VLSI devices used, the greater the learning time for the new devices and the greater the number of logic errors due to misapplication.

**Chips per Board.** Like many issues, that of how many chips should be squeezed onto a board requires a balance. The extremes of too many or too few are both bad. On one hand, too many chips on a board are bad because:

- Heat dissipation increases. This can put a strain on cooling and reduces reliability.

- The cost of the baseboard increases because there are more signal paths to provide.

- Cramped chip layout can increase the manufacturing and debug time per chip.

- There are usually few spare IC locations, which reduces upgradability and the ease of making changes.

- Board replacement as a repair technique becomes more expensive. Assuming no change in the failure rate, more chips per board gives the same number of failed boards, but each is more expensive to repair because it is more densely packed.

On the other hand, too few chips per board causes:

- More boards per system. They are each simple, but there are many of them. This increases both cost and weight.

- Chassis volume must be increased to hold all the boards.

- Inter-board communication is increased. This increases the system integration effort if the communication paths are manually installed, such as cable or wire-wrapped wires. The typically increased length and reduced noise immunity of board-to-board communication vs. on-board communication either reduces reliability or increases cost.

Weigh these opposing demands to find the proper density, and remember the demands of modularity. Each board should ideally hold modules that are strongly related with each other but less strongly related to those on other boards.

**Power Requirements.** Early in the design phase (perhaps after the first pass of the list of materials is made), the power requirements for each voltage supply should be estimated. This way, the power supply and cooling requirements can be considered in parallel with the board design.

**Use Second-Sourced, Available Parts.** Competition provides a better price and superior availability than found with a sole source, and there is no worry of a solitary manufacturer going out of business.

Availability is also important. Many delivery dates for new parts have been missed. Be sure the schedule conservatively reflects lead times for parts. Try to get samples of new parts before committing to use them. Speeds of new devices sometimes change from the preliminary specification to the final specification (hence the label "preliminary"). If the speed is important for the design, try to get some assurance from the manufacturer that the specification is firm.

**Minimize Number of Different Devices.** Of course, if a certain integrated circuit is the best for the job, it must be used. However, many different kinds of registers exist, shift registers, counters, and other popular MSI and SSI devices. Do not use synonymous parts unless they are necessary; by reducing the number of different devices, the likelihood of obtaining all parts on time is increased (even "available" parts can take months to arrive). Purchasing and incoming inspection personnel also appreciate a smaller list of materials. From the designer's standpoint, there are fewer devices that must be intimately understood, and there is less chance of confusion between similar but not identical devices.

**Board Layout.** Like design, the layout of chips on a board is a top-down process. The logic diagrams are already partitioned according to function, and, typically, all the devices relating to one function should be close together. The first step is to approximate the number of chips in each section (to find the board area each requires) and to mentally move these different-sized sections around. Be sure to put those sections pertaining to the backplane as close as possible to the edge connectors (an extra inch or two between drivers/receivers and the backplane can lead to problems, if a backplane signal is daisy-chained to many of these boards). Another constraint to consider is the cooling of the board: does the chassis cool any areas of the board better than others? If so, try to position the hotter sections accordingly. Orient the SIPs and the fins of heat sink so they are parallel to the direction of air flow.

Draw the sections to scale on an outline of the board. Allow the data flow to guide the placement — when done, you should be able to imagine data flowing in a pleasing path from section to section. If you find the data being shipped back and forth across the board in an inefficient fashion, there may be a better layout.

With this coarse layout done, the final placement for the chips in each section can proceed. If possible, keep all chips (except, perhaps, some large ones) oriented the same way and in regular rows and columns — this simplifies manufacturing, debugging, and testing.

To maximize the number of chips on a board, try as little as 0.1″ side-to-side spacing for the chips, and put blocks of chips of the same size (dynamic RAMs, for example) in a compressed format. Of course, if not absolutely necessary, do not crowd chips — this complicates manufacturing and testing. The placement of spares is important. Anticipate changes and put the spares in the areas where they are most likely to be needed. Make sure you have enough. A reasonable goal for the number of spares is at least 10% of the number of significant chips (exclude trivial blocks of chips like RAMs and PROMs in the chip count). Expect to spend several hours to several days evaluating all the constraints and finalizing the placement of the chips on the board. Board layout should be performed by (or, at least, approved by) the engineer, not the drafting department.

## 4.2 Low-Level Manufacturing Considerations

Some low-level mechanical, electrical, and practical aspects of manufacturing are discussed here.

 **IC Packaging.** Many integrated circuit packaging techniques exist. The following list covers the most common:

- **Dual in-line package (DIP).** DIPs are the most common packaging technique. The silicon device is packaged in a rectangular plastic or ceramic case and two rows of leads provide access to internal signals. The DIP leads are perpendicular to the plane of the case and are inserted into a socket or through a board. A ceramic case provides better heat dissipation but usually costs more.

- **Single in-line package (SIP).** SIPs have only one set of leads, and they lie in the plane of the case. SIPs have a smaller footprint (amount of board area consumed) than DIPs but often raise the board profile.

- **Flatpacks.** The flatpack has a smaller footprint than the DIP. It is a square device with flat leads on all four sides lying in the plane of the case, and it is mounted on the board surface. Its smaller leads do not reduce signal speeds as much as those of the DIP.

- **Small outline (SO).** A surface-mounted device, the SO package looks like a tiny DIP. It provides advantages similar to those of the flatpack. This packaging technique is rapidly increasing in popularity and is expected to be second to the DIP in sales volume before 1990.

- **Leadless chip carrier (LCC).** LCCs look like flatpacks without the leads. These are often used with sockets and are expected to be the third most popular package before 1990.

- **Pin grid array (PGA).** The PGA is a square case with many rows of pins underneath and perpendicular to the case. The PGA and its variants offer the highest number of signal pins per unit area, an increasingly important trait as VLSI devices demand more signals per device.

As board densities become higher, the trend away from DIPs to the newer techniques will increase. Unfortunately, the smaller case areas can cause heat problems. The smaller techniques also have difficulty competing with DIPs in the areas of cost and ease of debugging (the small leads are more difficult to access) [2, 3]. (For more information, see Date Codes in Section 5.3.)

**Grounded Signals.** Never ground signals by attaching pins to the ground plane; use a wire or trace. Not only are pins attached to the ground plane through a printed circuit (PC) trace difficult to change, but they are additional places to hand solder for a wire-wrap board.

**Thermal Relief.** Power and ground connections from chip baseboard holes to the power or ground planes are often made as massive as possible to provide good electrical contact. Unfortunately, this large contact area between the hole and the power or ground plane causes manufacturing difficulties by absorbing the heat required to properly draw molten solder into the hole and make a good connection. A compromise is required: by connecting the hole to the power/ground plane at only a few points, both electrical and manufacturing requirements are well met (see Figure 4.2).

**Silkscreen.** The silkscreen defines the inking of row letters, column numbers, text, and so forth on the board. The silkscreen is especially valuable on a prototype baseboard, which may have many man-months of effort expended on it, but it also helps test technicians and field technicians debug production boards. Consider the following for possible inclusion in the silkscreen:

- row letters on both left and right sides

- column numbers on both top and bottom

- horizontal lines connecting the top pins of all chips in a row

- boxes delimiting the three to ten major areas on the board (use a different line thickness or line texture from the horizontal lines)

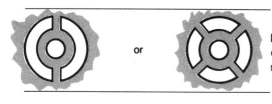

**Figure 4.2** Thermal Relief Proverb: Baseboard power/ground connection providing thermal relief.

- a "+" at the positive side of electrolytic capacitors (connecting this capacitor backwards destroys it). Clearly label the holes for other components which must be inserted a certain way.

- wire list and silkscreen revision level information

- a dot near pin 1 of each device (for orientation)

- labelling of the edge connectors (for answering the questions: Which connector is this? Where is pin 1? Which side of the board has the odd pins?)

- a slash every 10 edge connector pins for easy counting.

Silkscreen the chosen information on both sides of the board, especially the prototype board.

Consider the orientation of the silkscreen. The text across the top edge (purpose of the board, company name, part number, and so forth) should be readable when the cards are in the chassis. Define the debugging/testing setup before the orientation of the rest of the text (row letters, column numbers, and so forth) is determined. Each part of the silkscreen should be oriented so that it is readable by its audience.

Add appropriate warnings when applicable. For example, warnings might be placed near a hot heat sink or a high voltage source. Choose a distinctive ink color when possible and a legible lettering height.

Optionally, put the device number (e.g., "74S74") under each chip, surround the device number with a box to show the number of pins in the device, and label each linear device. This is bothersome in that the silkscreen must be updated after component changes, but it benefits manufacturing personnel. Another optional addition: a short phrase, oriented and placed for easy viewing, describing the board in more depth than the part number (e.g., "1MB Memory" or "Processor" or "Winchester Disk Controller"). (For more information, see Chip Location Labelling in Section 4.1 and Row Letters in Section 5.3.)

**On/Off Labelling.** A popular technique for labelling on/off switches uses a numeral "1" for on and a "0" for off instead of the words "on" and "off." While this may seem to be a step backwards in clarity, it proves to be a more universally-understandable scheme to nonEnglish-speaking users. Also, 1 and 0 are more dissimilar in appearance than "on" and "off." Other on/off indicators are 1) green for on and red for off; and 2) up for on and down for off.

**Spare Chip Locations.** Consider double rows of holes at spare board locations. The addition of a new chip is simplified by providing one set of holes for the DIP pins and another for connecting wires (see Figure 4.3).

**Device Profiles.** Beware of a few tall devices: SIPs, switches, ribbon cable connectors, linear devices, heat sinks, or perhaps devices in sockets. These can cause hot spots by blocking air flow to the neighboring devices. A board resting on the component side or with other boards laying on top of it has considerable force applied to its highest-profile devices. Keep in mind that inserting boards into chassis becomes difficult as boards get thicker.

The pitch of a chassis is the center-to-center spacing between boards. Common pitches for standard bus form factors are 0.6 inches and 0.8 inches. Given a pitch, the designer must subtract amounts for warping of the upper board, the length of pins protruding from the upper board, air clearance, and board thickness to arrive at the room remaining for device heights. (For more information, see Form Factor in Section 4.1.)

**Programmable Device Documentation.** PLAs, PROMs, and other programmable chips need a stick-on label to reference the documenta-

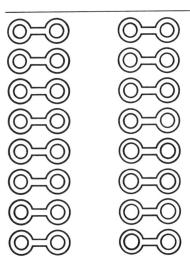

**Figure 4.3** Spare Chip Location Proverb: Spare chip location with extra row of holes for convenient wiring of a new chip.

tion that defines the contents. (For more information, see Programmable Logic and related proverbs in Section 2.4.)

**Chip Insertion.** Consider the requirements of automatic chip insertion equipment. If insertion is to be done by hand, understand the design demands, if any. Are there limits on the closeness of chips? Are DIP capacitors easier or harder to insert than radial-lead (cylindrical) capacitors? What must the silkscreen contain to simplify manufacturing?

**Placement of Changeable Devices.** If convenient, place DIP switches, devices in sockets (PROMs, crystals, and so forth), as well as other devices that might frequently change as close as possible to the top edge of the board. The top edge is the edge farthest from the backplane. Board modifications can be made without completely removing the board from the chassis. Some switches have their toggles on the side. If they are positioned properly, they can be changed without removing the board at all. (However, if changing settings with the power on can cause damage, perhaps these switches should be purposely recessed.)

**Status Displays.** Consider placing LEDs on the top edge of the board. These could show status and allow the user to tell at a glance if operation is correct. Tiny LCD digit displays are also available.

**Keyed Connectors.** Use cable connectors that fit together in only one way. This prevents incorrect connections.

**Signal Taps.** A valuable board-testing technique is the provision of labelled signal taps. The signal taps are short posts onto which an oscilloscope probe can be easily attached. Combine them with a clear and thorough testing procedure. Not only does this benefit test and field technicians, but also someone inexperienced in troubleshooting the board (such as a customer) may be able to make rudimentary measurements to determine whether the board works. Watch out for possible impedance mismatches caused by the taps. (For more information, see Impedance Mismatching in Section 3.3).

**Sleeving.** Put an insulating sleeve over long component leads where contact with other leads is possible.

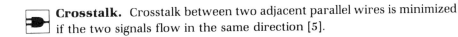

**Bends in Signals.** Wires in a printed circuit or Multiwire board should not have abrupt bends; this reduces the line's characteristic impedance that can cause reflections [5, 6].

**Crosstalk.** Crosstalk between two adjacent parallel wires is minimized if the two signals flow in the same direction [5].

## References

[1] Blakeslee, T. R., *Digital Design with Standard MSI and LSI* (John Wiley & Sons, 1979), pp. 17–19.

[2] Balde, J. W. and Brown, D., "Alternatives in VLSI Packaging," *VLSI Design*, December 1983, pp. 23–29.

[3] Stubbs, G., ed., "Surface-Mount SO Packages to Lead IC-Package Growth," *EDN*, September 20, 1984, p. 406.

[4] *Integrated Fuse Logic Data Manual 1984* (Sunnyvale: Signetics Corp., 1984), p. 2–3.

[5] Williams, M. and Miller, S., *Series 54ALS/74ALS Schottky TTL Applications* (Dallas: Texas Instruments Inc., #B215), p. 16.

[6] Blood, W. R. et al., *MECL System Design Handbook*, Revised Edition (Phoenix, AZ: Motorola Inc., 1980), p. 31.

[7] Donn, E., "Estimate Prices for Electronic Equipment Merely by Counting the ICs," *Electronic Design*, September 27, 1979, p. 92.

# 5
## Debugging Rules

Chapter 5 discusses the debugging of prototypes, but is is also applicable to the testing of production boards.

## 5.1 High-Level Philosophy

**Debugging Philosophy.** Debug by modules, going from the most independent functions (those that take very few inputs) to the most dependent. In choosing the module debugging order, consider which modules could most help the debug of other modules. For example, having a working system interface might be an excellent first step. Try to debug each function thoroughly before going to the next; it is valuable to know that the inputs coming from the debugged modules are correct.

Errors can be grouped into three classes:

1. Logic/conceptual errors. The algorithm is incorrect.

2. Noise, electrical, and physical errors. These can be difficult to find and are due to incomplete consideration of the analog traits of design. These traits include propagation delay, reflections, signal length, fanout, noise, decoupling, termination, and heat.

3. Manufacturing errors. These problems include inadequate soldering, wiring errors, and failed devices.

Any technique that focuses on only certain failure classes can help. Knowing that an error class has been eliminated from consideration simplifies finding the problem.

If sockets are being used, one debugging strategy is to put in components only for the section of the board being debugged. As each new section's chips are added, any errors in previously working hardware are almost surely caused by the new devices. Finding this type of error with all chips in place is considerably more difficult.

Regression testing retests the previously working hardware to ensure continued correct operation. Devise a confidence test for each debugged module so that it can be quickly regression tested after modifications are made. Preliminary documentation may be valuable for debug assistance. (See Debugging Tools in Section 1.3, Slower Crystal in Section 5.3, and Board Communication in Section 2.2 for some suggested design enhancements which assist debugging.)

**Error Grouping.** Errors tend to occur in groups — they are not evenly distributed. If a section has produced a unusually large percentage of the errors discovered, it probably contains a disproportionately large share of the *remaining* errors [1].

Why is this so? That errors usually occur in groups is not difficult to see, especially when some sections are new or challenging and others simple variations on previous work. A single misunderstanding often leads to several errors. This usually occurs in a challenging section, which further broadens the gaps between the numbers of errors in sections. Even in the event that all sections are of equal complexity, imagine the unlikeliness of each section having an equal number of errors.

Why does the distribution of discovered errors predict the distribution of errors remaining? Let's imagine a related situation in which a gumball machine, when requested, dispenses a colored gumball (analogous to an error); the color is either red or blue (the error is either in section A or section B). Suppose further that the machine has a finite supply of gumballs and dispenses them randomly, but because the globe holding the balls is opaque, we cannot see the colors of the remaining gumballs. Knowing the colors of the gumballs we have already received, we can guess at the colors remaining (assuming that they are randomly received). The distribution of remaining gumballs between red and blue is likely close to the distribution of the gumballs received. Of course, the more gumballs received, the more accurate this correlation. By the way, this process of guessing at the characteristics of many items from a small sample is used in calculating the mean time between failure (MTBF) [7].

Thinking of the gumballs as errors, we see that the observed distribution of errors over various modules gives a strong clue to the distribution of errors waiting to be found. The assumption that the errors are found randomly is fundamental to the validity of the gumball analogy; in practice, your debugging strategy may detect errors in sections randomly or may focus on each module in turn and debug it exhaustively.

**Corrections.** Corrections are more error-prone than the remaining hardware [2]. Keep this in mind especially during the end of the debug phase and into production. In this part of a product's life cycle, the rate of error discovery can be (and hopefully is) quite low. Without an engineer's full attention to the board, the broad understanding of it is lost. Fixes are plausible on the surface, but they are often done with a lack of full understanding of the problem and of the board itself. This is why fixes are error prone and why old fixes should often be re-examined first when analyzing a new problem.

Be sure to fully document corrections to a project, *especially* if more than one prototype is being built concurrently. Avoid the waste of rediscovering a problem (see Debugging Notebook Format in Section 2.3).

One final thought: understand the problem completely before devising a solution. Be sure that you are solving the problem, not just a symptom.

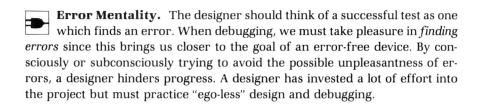

**Error Mentality.** The designer should think of a successful test as one which finds an error. When debugging, we must take pleasure in *finding errors* since this brings us closer to the goal of an error-free device. By consciously or subconsciously trying to avoid the possible unpleasantness of errors, a designer hinders progress. A designer has invested a lot of effort into the project but must practice "ego-less" design and debugging.

**Assistant Debugger.** The designer may have difficulty in forming the "devil's advocate" mentality valuable for exposing errors. Also, if the designer misunderstood a part of the design specification, there is difficulty in spotting problems with that part. Someone who did *not* participate in the design might be especially useful in devising tests.

**Test Sequence.** Plan a test sequence so that debugging time is efficiently spent. Keep the available test equipment in mind and plan (early in the design) which equipment must be procured for debugging.

**Test Approach.** First, define the expected output of a test (perhaps with timing diagrams, patterns on a monitor, or values in a memory) and *then* execute the test. Otherwise, one might interpret the results as correct without fully understanding what correct results should look like. "If there are no expectations, there can be no surprises" [3].

**Change One Thing at a Time.** Be pessimistic, and change only one thing at a time. If several things are changed before the project is rechecked and one does not work, the debugger has little idea where to look.

**Test Everything.** A design often contains hardware to handle rare situations such as invalid inputs, special cases or border conditions, or perhaps the failure of a peripheral. Do not forget about this hardware in preparing test input. The alternative, of course, is a "completed" design with rarely-invoked but essential hardware either untested or incompletely tested.

## 5.2 Design Reviews

Design reviews form an important debugging tool. At various milestones in the project, the design should be objectively reviewed by a small group of the engineer's peers and anyone else who can contribute [8]. The goal is to dig into the design — not the designer.

**Preparation.** Give reviewers at least a week to look over your logics, timing diagrams, and other documentation. They should prepare preliminary comments and questions. The more effort invested by the designer and the review team, the more productive the review. To focus the discussion, separate reviews might be conducted. Engineering reviewers might concentrate on the design, while marketing, manufacturing, and other reviewers might look at the design from other angles in a separate review.

A moderator (*not* the designer) should be assigned prior to the review date. During the review, the moderator should keep order, see that all are heard, prevent digression, and maintain the proper level of detail.

**Find Errors.** In design reviews, the designer must not take errors found personally and must encourage the reviewers to find errors. Assuming that there are errors in the design, a design review that discovers no errors is a waste of time. The more errors found, the less remain and the better — not worse — the designer should feel (see Error Mentality in Section 5.1).

**Raise Issues; Don't Resolve.** In a design review, reviewers should raise issues without trying to resolve them. The designer is obliged to respond to the issues some short time after the review. Discussing solutions to problems at too low a level during the review reduces the breadth of coverage that a review team can give a design.

## 5.3 Low-Level Considerations

**Slower Crystal.** Whenever possible, put a slower crystal in at the beginning of the project. With a slower clock, you can focus on logic errors leaving propagation, noise, and other device-dependent problems for later. A large frequency change cannot be tolerated by devices that are linked to an externally-defined frequency, such as a video refresh frequency or that have chips with a minimum clock rate, such as a microprocessor or dynamic RAM refresh circuit. However, a small change may be all that is needed.

Anticipating single step or reduced frequency operation, try to avoid placing constraints on the clock speed. For example, keep an asynchronous interface really asynchronous. Do not assume that the time between event A and event B is, say, three of our clock periods. This type of assumption ties together our board and either another board with which we communicate or a bus. If our board's interface changes — because of debug at a different frequency, design corrections, or updates — the interface fails (see also Minimize Coupling in Section 2.1).

**Single Stepping.** For those devices not linked to an externally-defined frequency (such as the frequency of a video device or other peripheral), consider a single-step circuit with a pushbutton. Make sure that no device has a minimum clock frequency (such as a processor or DRAM refresh circuit).

**Video Applications.** In video applications, try to get *something* on the screen as soon as possible. The monitor can be a very useful debugging tool by creating visual clues to errors.

**Row Letters.** Defining row letters and column numbers is one way to identify chip locations. To specify a single pin with this layout, the column of the chip, row of the chip, and the pin number must be identified. A convenient way of specifying pin 5 on the chip in column 20 and row A would be: "20A5."

The letters on the rows of chips should be chosen so that they *sound* different (i.e., don't rhyme). This can avoid confusion when two debuggers work together and verbally refer to board locations. This also simplifies the manual creation of a wire list, where one person often reads locations to another. Avoiding the letters I and O (which can look like 1 and 0) and all the redundant letters in the groups that sound alike, the possible row designations are: A, B, F, H, L, M, Q, R, S, W, X, and Y. Additional possibilities are short words or letters from other alphabets (see Chip Location Labelling in Section 4.1 and Silkscreen in Section 4.2).

**Expensive Chips.** Put in expensive or hard-to-get devices, such as manufacturers' samples, *after* their control inputs and power/ground connections have been examined so these valuable devices will not be damaged.

**Acceptance Test Procedure (ATP).** Adequately documenting all problem symptoms and solutions as they are found assists in the writing of a software ATP at the end of the project. Consider having tests to uncover and pinpoint all or most of the errors found during the course of debugging. Also, try to make the ATP require as little user interaction as possible (such as viewing a monitor or making other visual checks) to maximize automation. Those rare boards that cannot interact with a computer need a more hands-on ATP.

**Document Secondary Problems.** Document problems found during the course of solving another problem. Often, a designer regrets not having paid more attention to a problem when it was first noticed. Note symptoms for later investigation, but avoid getting sidetracked. Try to completely analyze one problem before attacking another. Often, solving one problem unexpectedly eliminates seemingly unrelated symptoms.

**Debugging Notebook Format.** A notebook is valuable for recording the project's progress. With several engineers working at different times, it provides a good communication tool. Also, recording the symptoms of a problem and its solution are valuable when the problem occurs again. In the debugging notebook, use the following (or similar) format for all problems:

```
        SYMPTOMS:  ....  (what the hardware was doing that was wrong)
REASON (or PROBLEM):  ....  (what error had been made)
        SOLUTION:  ....  (what was done to fix the problem)
```

For example,

```
SYMPTOMS:  The ALU at location 21B (page 10 in the logics) was
           doing this wrong:  ...  under these circumstances:  ....
  REASON:  The ALU's carry-in input was left floating
SOLUTION:  Wire 20B16 to 21B7 (connect the carry-out from the
           low 4 bits to the carry-in of the upper 4 bits).
```

Be sure to date each entry and record the status of the solution (has it been implemented yet?).

Some reflections on why the problem occurred and how it could be prevented from happening next time are also valuable. Problems often recur, hence the importance of thoroughly documenting the problems the first time.

**Source of Error.** After an error is discovered, an analysis of how it occurred may slightly help the schedule of the current project but can greatly educate the people responsible for the error. The answers to the following questions might be documented (adapted from Myers [6]):

- **When was the error made?** Was it an ambiguous statement in the specifications? a logic design flaw? or was made an hour ago with an improperly-thought-out fix to some nearby section?

- **What was done incorrectly?** Was it the failure to draw a timing diagram? too little time spent in defining the specification? an invalid assumption? the failure to consider all inputs? Many errors are neither logical nor electrical in nature — for example, a chip failure or an incorrectly placed wire.

- **How could the error have been prevented?** What should be done next time to prevent this type of error?

- **How could the error have been detected earlier?** If we cannot prevent this error, how can the review and testing process be improved to detect it earlier? An error detected at a certain design stage is many times more costly than the same error detected during the previous stage. Consider the difference in effort of fixing an error during design time (involving logic redrawing) vs. after the product is in the field (modifying formalized documentation and correcting devices). Fleckenstein [4] compares effort and expense in implementing *software* modifications at various stages during a project. The conclusion was that field fixes were one hundred times more expensive than fixes made to the specification. The ratio is likely to be higher for hardware.

- **How was the error found?** Something valuable has happened: we have found an error. What technique did we use to find this error? Can we learn something from this? Might this technique find more errors in the current project?

- **How long did it take to find this error?** If we group the time to find an error into class 1 (one hour or less), class 2 (up to four hours), or class 3 (over four hours), we know the relative importance of what we have learned. This might also be valuable for statistics evaluated at the end of the project.

**CAUTION:** This log of errors is for the use of the project team and *not* management. It can be an excellent procedure for self-improvement. Using it for employee review, however, defeats its purpose and discourages its frank and open use (see Design Reviews in Section 5.2).

**Impedance Mismatch in Practice.** If the load impedance is higher than the source impedance, inductive ringing is likely, due to the underdamped nature of the line (this is what typically happens with TTL and ECL). On the other hand, a higher source impedance than the load impedance can cause capacitive charging, because the line is overdamped.

A powerful clue that the signal has a problem with reflection is that the signal appears differently at different places on the line, due to the interaction of the reflections bouncing off the ends (5) (see Impedance Mismatching in Section 3.3).

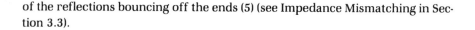

**Shaky Waveforms.** If the oscilloscope shows an edge in a supposedly periodic waveform to be jittery, this edge is clearly aperiodic with respect to the scope trigger. Metastability is the possible cause if the edges change state more slowly than normal and look somewhat like rain on the oscilloscope. The many different waveform edges typically become stable within a certain time window ending 10–100 nanoseconds after the state change should have completed (see also Metastability in Section 2.4).

**Freeze Spray.** Freon sprays are available to cool chips. They can be valuable in determining how fast a hot chip will heat, after being cooled, and in changing the propagation delay of a device. The propagation delay of a 7400 TTL device is usually fastest at room temperature (see Propagation Delay vs. Temperature in Section 3.1). However, a rapid temperature change can cause sufficient stress to create a failure.

Don't forget that the commercial temperature range usually only drops to 0° Centigrade, operationally, but that the sprays can cool to −50° Centigrade. Watch out for excessive water condensation, too.

**Thermometers.** Stick-on thermometers (either permanent or reversible) can be valuable for determining how close a chip's temperature comes to its upper limit. The permanent kind is especially useful in recording the chip temperature on boards inside a chassis, which cannot be seen while running. Reversible thermometers always track the current temperature while permanent ones hold the peak temperature. (See Junction Temperature in Section 3.1).

**Date Codes.** Certain batches of chips can have a much higher-than-normal failure rate due either to manufacturing problems or improper handling. Watch for a link between a group of failures of a part and a certain date code; if a batch of chips is bad, either replace them or be aware of their increased error proneness. The date code is printed on a chip along with the device number. (For more information, see Device Failure Rates in Section 3.1.)

**Safety.** Power supplies used with digital devices often supply tens and even hundreds of amps direct current (DC). It is vital that this power be treated with respect. The severity of a shock is usually proportional to the current that flows through the body. The voltage of the electricity determines how much current can pass through the skin. The skin normally has a fairly high resistance, but the resistance can be lowered with damp hands or feet or raised with insulating gloves or shoes. DC is usually safer, requiring several times the current to produce the effect of a 60 Hz alternating current (AC) shock. The path taken by the current through the body is also important. Current through the chest (perhaps going from one hand to the other) is especially dangerous. For this reason, use insulated tools and try to work with only one hand, especially when performing unfamiliar or dangerous electrical tasks.

Grounded (3-way) plugs increase safety, but make sure that the power outlet is properly wired. Watch out for water and damp hands and remove or cover rings, metal watchbands, and other conductive objects that might make a power-to-ground connection. Even 5-volt power supplies can be *very* dangerous if the body's impedance is lowered.

Some European safety standards require that a system's on/off switch turns off *everything*. Devices following these rules are no different whether turned off or unplugged. This eliminates, for example, a small amount of power being used to maintain the contents of RAMs or to power a clock chip.

**Debug Power and Ground.** Before applying power to a new prototype board, ensure that there is no connection between the power and ground inputs (and between all the power inputs, if there are more than one). It is easy to accidentally connect power and ground together either with a wire on a wire-wrap board, or with a copper trace or solder bridge on a printed circuit board or a wire-wrap baseboard. If such a connection is found, try to locate the trouble with a milli-Ohmmeter. This device measures the small resistance of the wires or baseboard traces and, as you probe around the board, reports when you have shortened the path. This method quickly pinpoints the source of the problem (see Board Manufacturing Techniques in Section 4.1).

**Floating Inputs.** TTL inputs float more or less high, but are susceptible to noise if floating. Use a pullup or ground to provide a solid constant input. ECL, on the other hand, floats to a solid, reliable low (see Pullups in Section 3.2).

## References

[1] Myers, G. J., *The Art of Software Testing* (John Wiley & Sons, 1979), pp. 15–16.

[2] Ibid., pp. 143–144.

[3] Ibid., p. 12.

[4] Fleckenstein, W. O., "Challenges in Software Development," *Computer*, March 1983, pp. 60–64.

[5] Davidson, M. F., "Understanding the High Speed Digital Logic Signal," *Computer Design*, November 1982, pp. 79–82.

[6] Myers, pp. 144–146.

[7] Blakeslee, T. R., *Digital Design with Standard MSI and LSI* (John Wiley & Sons, 1979) pp. 350–351.

[8] Myers, pp. 17–35.

# 6
## Documentation Rules

Perhaps documentation rules is the area where opinion becomes most necessary and where the least clear distinction between "the correct way" and other ways exists. These rules are intended only as suggestions and should be tempered by your experience, your preference, and the application at hand.

## 6.1 High-Level Philosophy

 **Modularize.** Timing diagrams and logic drawings should be modular. The split of logic onto the pages affects design phases from logic design and board layout through documentation. It should be carefully thought out. Having to resegment the logics at project completion is very time consuming and error prone; all other documentation done to that point must be modified to reflect the new layout. Where convenient, put timing diagrams next to the related logic they describe.

**Bus Labelling.** Carefully consider whether the lines in a bus should be labelled starting at 0 or at 1 for the least significant bit. Any busses that can be considered together to represent a single binary number (e.g., an address bus or data bus) should be numbered starting at 0. This simplifies the mental conversion of the values on a bus into a binary number. All other busses can be numbered starting at 1, if desired, although consistency might demand a uniform numbering scheme.

**Design Languages.** Several hardware design languages exist. These include:

- CDL (Computer Design Language) describes low-level hardware execution sequences

- ISP (Instruction Set Processor) can describe all the storage available to a programmer (cognition codes, index registers, addressing modes, and so forth)

- AHPL (A Hardware Programming Language) clearly and completely defines algorithms in an APL-like format (APL is a computer language)

- PMS describes a system in block diagram form

A design language can document structure well. Some languages also simulate the design. Simulation can be a very efficient technique for finding bugs. The use of a standard language (such as those above) instead of a custom language might be valuable in that the standard language has withstood the test of time and is more widely known [2,7].

**Clarity and Maintainability.** The logic drawings should be neat, organized, properly partitioned, and easy to understand. The placement of the chips should be defined (or, at least, approved) by the designer, not the drafting department. (For more information, see Board Layout in Section 4.1.)

## 6.2 Low-Level Drafting Details

The set of logic drafting rules should be tempered by your opinions. However, whatever set of rules you use, follow them *consistently*.

**Block Diagram.** The overall block diagram is an essential part of the logics. It should be one of the first pages in the logics (and should be the first page drawn).

For complex projects, a hierarchy of block diagrams may be necessary. Use only one page for the top-level block diagram. If the conceptual jump from a block to the logics that implement it is too great, expand that block with a block diagram of its own. The logics form the last level in the hierarchy.

**Timing Diagrams.** Consider the order of signals in the timing diagram. Inputs to the module are always first, outputs always last. External signals, not in this module, may be included to show the phase relationship between the two modules. They are placed before the input signals. For synchronous circuits, always include the master clock or a derivative for reference. Put down neither too many nor too few signals.

Timing diagrams are part of the documentation. Try putting them in an empty area on the logic diagram to which they refer. Watch the level of detail: too much and too little are both bad. Where it helps, draw arrows from signal levels or edges to the levels they cause. Where glitches are possible (for example, in the decoding of a counter output), draw them as vertical lines. Manufacturer's data books are often good sources of example timing diagrams. (For more information, see Use Top-Down Design in Section 2.1.)

**Arrows in Logics.** Use arrows on signals travelling left or up (horizontal and vertical signal lines without arrows are assumed to be travelling right and down). Periodically, bidirectional busses should have arrows going both directions. Arrows help emphasize the direction of the source.

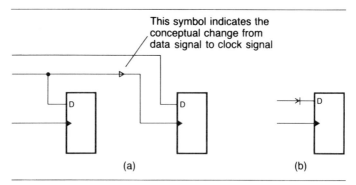

**Figure 6.1** Clock Transformation Proverb: Clock/data transformation documentation. (a) A signal intended for use primarily as data being used as a clock has the data-to-clock symbol. (b) Clock-to-data symbol.

**Clock Transformation.** Use the symbols in Figure 6.1 to represent data-to-clock and clock-to-data transformations. These symbols should appear over the signal at the point where the meaning changes. It helps to make the distinction between signals used as clocks and those used as data. It can be dangerous to casually use data values as clocks in synchronous design.

**Lettering Size.** Make the lettering large enough. Lettering and the spacing between lines that are too fine for the sheet size can quickly become illegible when the logics are reduced. Table 6.1 suggests some minimums.

| Sheet Size | Dimensions (inches) | Minimum Line Spacing (inches) | Minimum Lettering Height (inches) |
| --- | --- | --- | --- |
| A | 8.5 x 11 | 0.088 | 0.044 |
| B | 11 x 17 | 1/8 = 0.125 | 1/16 = 0.63 |
| C | 17 x 22 | 0.177 | 0.088 |
| D | 22 x 34 | 1/4 = 0.25 | 1/8 = 0.125 |

**Table 6.1** Lettering and line spacing requirements.

**Off-Page References.** Be sure to put the page numbers of the logics from which input signals come and to where output signals go. Put the page numbers in parentheses near the signal label. For busses, pick one source page, and let it be *the* source for the purposes of page-to-page references.

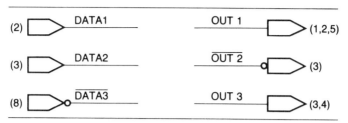

**Figure 6.2** Off-Page References Proverb: Off-page references to signals.

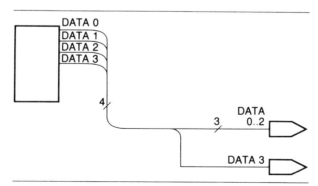**Drawing Busses.** Busses (groups of two or more signals that perform one function) should bend in a curve, not at a right angle. Inputs to and outputs from the bus also merge with the bus with a curve. Label all signals that join with or break away from the bus (see Figure 6.3).

**Constant Inputs.** Any input tied to a constant voltage (low or high) should have parentheses around the input name, which is inside the chip. By downplaying the visual importance of constant inputs, variable inputs are highlighted.

**Logic Symbology.** Use the new ANSI/IEEE standard logic symbology, Y32.14 [1, 3]. This style provides many benefits over more casual and less informative symbologies and has become standard with several manufacturers of TTL components (including Texas Instruments and Signetics). Y32.14 supersedes the Department of Defense (DOD) standard MIL-STD-806B. Devices

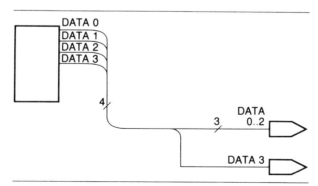

**Figure 6.3** Drawing Busses Proverb: Single signals merging with and splitting away from a bus.

in other families should also be drawn this way, even though the manufacturers may have their own techniques. Government agencies are increasingly specifying that work done for them be documented in this style (see Figures 1.5 and 6.4).

**Device I/O.** All inputs to and outputs from a device must be shown, whether they are used or not (exception: power/ground connections are *never* shown unless the device uses unusual voltages or nontraditional pins for its voltage supplies). Any no-connects should be labelled as such. No-connect pin numbers are also added. The input and output names for these no-connects, like those for constant inputs, are written in parentheses. Debugging and testing might need those unused inputs and outputs in obtaining more information. As examples of signals that might not be needed for the design but could aid debugging, consider Input Ready and Output Ready on a FIFO, Terminal Count on a counter, Carry Out on an ALU, or A = B on a comparator.

**Spare Devices.** List all unused gates, flip flops, and so forth. This list can save you time in looking for a spare logic element when implementing a correction or upgrade. Wire list programs, which maintain the list of wire connections, can automatically produce the list of unused logic elements.

**Logic Density.** Paper is cheap; do not crowd logics on too few pages. On the other hand, too many pages of logic increase page I/O and may be confusing.

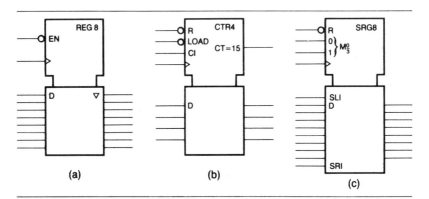

**Figure 6.4** Logic Symbology Proverb: Sample MSI devices in the new symbology: (a) register, (b) counter, and (c) shift register. The drawing style for gates is basically unchanged.

**Signal Labels.** Label as many signals as possible with meaningful names. This labelling is clearly essential on signals travelling between pages of logics, but is also valuable for signals internal to a page. The labels serve as comments and give a better idea of the use of each line.

Use illustrative names as signal labels, and do not be afraid to make the name long. Confusion due to a short, contracted name is easily avoided. Always put dashes in multi-word names. The only limit to the name's length is that imposed by some simulators or other software packages (see $\overline{RAS}$ Signal for DRAMs in Section 3.1).

 **Comments.** Occasional comments are good ways to explain, in English, a complicated section. They may also make reference to a more detailed description, elsewhere. Keep the comments succinct. Names can also be added above devices to clarify their function (e.g., "ROW ADDRESS COUNTER").

 **Grouping.** To make the hierarchy obvious, make appropriate groupings of devices:

- A dashed rectangle with long dashes is used to enclose perhaps 5–20 devices which together perform a function. Put a name in the rectangle describing the function. Functions might be "STATE GENERATOR," "INSTRUCTION DECODER," or "BUS INTERFACE." A page of logics might have 2–10 of these. I/O between these rectangles should be minimized (to encourage information hiding). Put arrows on all I/O lines as they enter/leave each rectangle.

- A dashed rectangle with medium-length dashes encloses a very small number of devices (perhaps 1–4) that perform a single, simple function (e.g., 2 NAND gates making an S-R flip flop, a level-to-pulse converter, or any other function that might belong in Section 2.5, Helpful Circuits). The inputs and outputs are labelled to show the I/O of this virtual device.

- A dashed rectangle with small dashes encloses a group of devices that are in a single chip and are inseparable. These chips are best documented by drawing the several boxes or gates that form the function inside a single dashed rectangle. For example, devices that might be drawn this way are the 74S64 TTL AND-OR-INVERT combination, 74S51 AND-OR-INVERT, 10131 2-bit ECL flip flop with clock enable, PLA contents (if simple), and so forth.

 **Sampling.** Place the symbol shown in Figure 6.5 over a location where asynchronous sampling occurs. This is a warning for metastability.

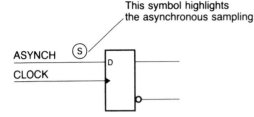

This symbol highlights the asynchronous sampling

ASYNCH

CLOCK

**Figure 6.5** Sampling Proverb: The flip flop samples asynchronously; its data input is labelled appropriately.

**Pullups and Grounds.** Pullups and grounds tie TTL signal inputs high and low, respectively. Label them as shown in Figure 6.6. With ECL, unattached inputs float to a solid low and grounds are used to create a constant high input. Draw an ECL ground as shown in Figure 6.6b except delete the inverting bubble.

**Chip Requirements.** In addition to the requirements of the ANSI/ IEEE standard symbology, every digital device needs:

- The board location in a circle, touching the drawing of the device (in a consistent corner)

- Every pin labelled and numbered (except power and ground)

- The device name inside the device (e.g., 74S151 or 74AS804).

**Title.** Give every page of logics a meaningful title. If a succinct title is not easily found, that may indicate that too much or too little is on the page. The page can be split with vertical or horizontal lines to make several narrower pages if the original page's shape is not optimal.

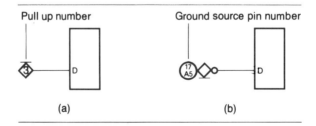

Pull up number

Ground source pin number

(a)

(b)

**Figure 6.6** Pullups and Grounds Proverb: (a) Pullup (the pullup number is in the diamond). (b) Ground (the ground source pin number is in the circle).

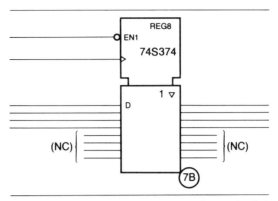

**Figure 6.7** Chip Requirements Proverb: Chip labelling requirements. Note the board location, the labelling of every pin (even those unused), and the device name.

**Upper-page Spacing.** Try to leave a blank 1.5 inches at the top of the logics so that they can be stapled there.

**Wiring Connections.** Connect wires with two "T" intersections instead of one "+" intersection. The ambiguity between nonconnecting and connecting crossovers is avoided (see Figure 6.8).

**Different Logic Families.** Draw a dashed line on the boundary between two logic families with different electrical characteristics (e.g., between TTL and ECL, but not between 74S and 74AS TTL). Put the names of the

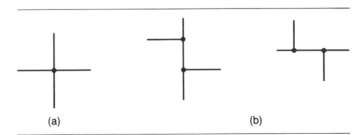

(a)                              (b)

**Figure 6.8** Wiring Connections Proverb: Wiring interconnections. (a) Incorrect. The reader may wonder whether this is a connecting or nonconnecting crossover. (b) Two correct replacements for the " + " interconnection. These cause no confusion.

families on the appropriate side of the dashed line so that it is obvious to a test engineer that different logic levels exist. The dashed line should go through the middle of translators. Not only does this inform the test engineer that high and low are interpreted differently, but it cautions that the chips may have different ground pins (to which oscilloscope ground leads are often attached). For more information, see Logic Families in Section 3.1.

**External Connections.** One page of the logics can be reserved for itemizing all external connections. These include backplane I/O and ribbon cable connections.

**Zone Coordinates.** Each drawing should have a scheme for referring to its subsections in the textual documentation. One way is to have letters on the left and right edges, and numbers across the top and bottom edges, defining a coarse grid. The text can easily and accurately refer to devices; for example, "The 4-bit counter in section C2 of page 5 is used to. . . ."

## 6.3 Textual Documentation

A strong and strengthening trend exists toward users demanding readable, informative documentation. The quality of documentation is becoming an increasingly important criterion in judging a product.

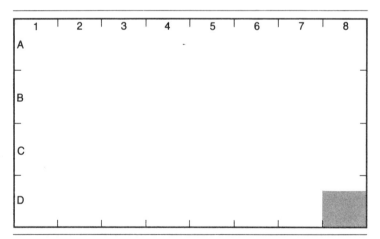

**Figure 6.9** Zone Coordinates Proverb: A sample drawing layout. The letters define rows and the numbers define columns, allowing a clear reference to a small region of the logic.

**Board I/O.** A user should not have to read all the documentation to learn about the input and output requirements of the board — a single, to-the-point section should describe it all. The discussion is brief, of course, with boards on a standard bus with no other external connections. Also include a summary of the settings of switches, jumpers, and other configuration-dependent devices. This section helps the user quickly and correctly install the board. (For more information, see Different Configurations in Section 2.2).

**Minimum Success Rate.** Consider a group of users obtaining their information about the use of a product *only* from the documentation (assume that they are not allowed any other input). Think of the readability and thoroughness of documentation in terms of the minimum acceptable success rate of this group.

Documentation should be thoroughly reviewed by persons with the same skills as a typical user, who are not familiar with the project. The intended audience and their assumed experience should be explained in the beginning of the document.

**Notes, Cautions, and Warnings.** Highlight especially important facts as notes, cautions, or warnings. These should have both margins indented and are defined as follows:

NOTE
An operating instruction or practice that should be emphasized.

CAUTION
An operating instruction or practice that, if not strictly observed, may result in damage to the equipment.

WARNING
An operating instruction or practice that, if not strictly observed, may result in personal injury.

For more information, see Safety in Section 5.3.

**Good Writing Practices.** Much has been written on good technical writing [4, 5, 6]. A few guidelines follow:

- Make an outline first, and follow it as you write. The introduction should touch on all major points in the outline.

- Use clear, common words. Replace "prior to" with "before," replace "utilize" with "use" replace "accomplished" with "finished," and replace "possesses" with "has."

- Eliminate deadwood. Replace "it should be noted that" with "note that" and "take into consideration" with "consider." Deadwood and the failure to use common words are among the most commonly broken writing rules. Keep writing clear and simple!

- Keep most sentences short and simple. Different sentence lengths help to maintain the reader's interest, but keep the average length somewhat short.

- Use the active voice. The sentence is clearer if the subject performs the action. For example, change "the trace must be cut by the technician" to "the technician must cut the trace" and change "the electron beam is controlled by the coils" to "the coils control the electron beam."

- Write as if you were speaking. Use a casual, conversational style. If the personal pronouns "I," "we," or "you" convey the meaning best, use them.

- Read your writing aloud. This often points out sections that are clumsy or confusing.

- Use headings and subheadings. These provide milestones to keep the reader on the right track and help with skimming and review. However, do not overdo them or make them too elaborate so that they detract.

- Do not be afraid to use numerals for numbers instead of words, even for numbers less than 10. Numerals are often much clearer. One exception: two numbers together can be confusing. This can be written clearly as in, for example, "three 8-inch wires."

- Figures and tables should have captions that are informative enough that the reader need not refer to text to understand them.

- Define acronyms, abbreviations, and "inside" terms when they first appear.

- Replace a long and tedious textual list with a table.

- Do not use a pronoun unless what it stands for (the antecedent) is clear.

- Repeat terms if needed; avoid synonyms. The reader may wonder if a synonym has different meaning than the original word.

- Avoid nouns created from verbs; use the verbs instead. Replace "specification" with "specify," replace "refusal" with "refuse," and replace "verification" with "verify."

- Avoid negatives. This is occasionally difficult, but try to change the construction of the sentence so that the "not" is replaced with a positive word. "Not" is a small word that is vital to the comprehension of the sentence but may be overlooked. For example, replace "do not force the board" with "avoid forcing the board."

- Use imperatives for instruction. Replace "the screw must be turned" with "turn the screw." This is a special case of the "Use active voice," above.

- Conditions must be easily understood. Replace a confusing condition such as "if A and B or C, then D" with a flowchart or a listing of the condition.

**American vs. European Numbering.** Documentation that is to be read outside of the U.S. should be written with some attention given to the regional differences in the writing of numbers. The number "12,345.6" is written as "12.345,6" in France (i.e., the roles of the comma and period are reversed). *Spectrum* magazine solves this problem by replacing the placeholder punctuation used to separate each three digit group with a space: "12 345.6."

"Million" is understood as $10^6$ in both the U.S. and Britain, but the British (and much of the rest of the world) consider a billion to be $10^{12}$. That is, instead of jumping three orders of magnitude from "million" to "billion" to "trillion," and so forth, some societies jump six orders of magnitude. A simple solution is to represent ambiguous numbers in scientific notation or to use metric prefixes such as "mega-," and "giga-."

The writing of dates with numbers also has several standards. The date "1-6-80" means "January 6, 1980" in the U.S., but "June 1, 1980" to most Europeans. An unambiguous compromise changes the order of the numbers to year-month-day; for example, January 6, 1980 becomes 80-1-6. Always writing the name of the month is clearest of all.

# References

[1]  *The TTL Data Book*, volume 1 (Dallas: Texas Instruments Inc., 1984), section 4.

[2]  Baer, J., *Computer Systems Architecture* (Computer Science Press, 1980), pp. 31–75.

[3]  *Graphic Symbols for Logic Functions* IEEE Std. 91-1984 (New York: IEEE, 1984).

[4] Felker, D. B., et al., *Guidelines for Document Designers* (Washington: American Institutes for Research, 1981).

[5] Eatherton, J., "Companies Crack Down on 'Wordy, Imprecise' Technical Writing Styles," *Electronic Engineering Times*, December 5, 1983, p. 107.

[6] Strunk, W. and White, E. B., *The Elements of Style*, Third Edition (Macmillan, 1979).

[7] December 1974 issue of *Computer*.

# Appendixes

## Appendix 1: Code and Number Table

| Decimal | Binary | Octal | Hex | Gray | BCD | ASCII | |
|---|---|---|---|---|---|---|---|
| 0 | 00000000 | 000 | 00 | 00000000 | 0000 | NUL | (Null) |
| 1 | 00000001 | 001 | 01 | 00000001 | 0001 | SOH | (Start of Heading) |
| 2 | 00000010 | 002 | 02 | 00000011 | 0010 | STX | (Start of Text) |
| 3 | 00000011 | 003 | 03 | 00000010 | 0011 | ETX | (End of Text) |
| 4 | 00000100 | 004 | 04 | 00000110 | 0100 | EOT | (End of Transmission) |
| 5 | 00000101 | 005 | 05 | 00000111 | 0101 | ENQ | (Enquire) |
| 6 | 00000110 | 006 | 06 | 00000101 | 0110 | ACK | (Acknowledge) |
| 7 | 00000111 | 007 | 07 | 00000100 | 0111 | BEL | (Bell) |
| 8 | 00001000 | 010 | 08 | 00001100 | 1000 | BS | (Backspace) |
| 9 | 00001001 | 011 | 09 | 00001101 | 1001 | HT | (Horizontal Tab) |
| 10 | 00001010 | 012 | 0A | 00001111 | — | LF | (Line Feed) |
| 11 | 00001011 | 013 | 0B | 00001110 | — | VT | (Vertical Tab) |
| 12 | 00001100 | 014 | 0C | 00001010 | — | FF | (Form Feed) |
| 13 | 00001101 | 015 | 0D | 00001011 | — | CR | (Carriage Return) |
| 14 | 00001110 | 016 | 0E | 00001001 | — | SO | (Shift Out) |
| 15 | 00001111 | 017 | 0F | 00001000 | — | SI | (Shift In) |
| 16 | 00010000 | 020 | 10 | 00011000 | — | DLE | (Data Link Escape) |
| 17 | 00010001 | 021 | 11 | 00011001 | — | DC1 | (XON, Device Control 1) |
| 18 | 00010010 | 022 | 12 | 00011011 | — | DC2 | (Device Control 2) |
| 19 | 00010011 | 023 | 13 | 00011010 | — | DC3 | (XOFF, Device Control 3) |
| 20 | 00010100 | 024 | 14 | 00011110 | — | DC4 | (Device Control 4) |
| 21 | 00010101 | 025 | 15 | 00011111 | — | NAK | (Negative Acknowledge) |
| 22 | 00010110 | 026 | 16 | 00011101 | — | SYN | (Synchronous Idle) |
| 23 | 00010111 | 027 | 17 | 00011100 | — | ETB | (End of Text Buffer) |
| 24 | 00011000 | 030 | 18 | 00010100 | — | CAN | (Cancel) |
| 25 | 00011001 | 031 | 19 | 00010101 | — | EM | (End of Medium) |
| 26 | 00011010 | 032 | 1A | 00010111 | — | SUB | (Substitute) |
| 27 | 00011011 | 033 | 1B | 00010110 | — | ESC | (Escape) |
| 28 | 00011100 | 034 | 1C | 00010010 | — | FS | (File Separator) |
| 29 | 00011101 | 035 | 1D | 00010011 | — | GS | (Group Separator) |
| 30 | 00011110 | 036 | 1E | 00010001 | — | RS | (Record Separator) |
| 31 | 00011111 | 037 | 1F | 00010000 | — | US | (Unit Separator) |
| 32 | 00100000 | 040 | 20 | 00110000 | — | SP | (Space) |
| 33 | 00100001 | 041 | 21 | 00110001 | — | ! | |
| 34 | 00100010 | 042 | 22 | 00110011 | — | " | |
| 35 | 00100011 | 043 | 23 | 00110010 | — | # | |
| 36 | 00100100 | 044 | 24 | 00110110 | — | $ | |

## Appendix 1 (*cont.*)

| Decimal | Binary | Octal | Hex | Gray | BCD | ASCII | |
|---|---|---|---|---|---|---|---|
| 37 | 00100101 | 045 | 25 | 00110111 | — | % | |
| 38 | 00100110 | 046 | 26 | 00110101 | — | & | |
| 39 | 00100111 | 047 | 27 | 00110100 | — | ' | (Apostrophe) |
| 40 | 00101000 | 050 | 28 | 00111100 | — | ( | |
| 41 | 00101001 | 051 | 29 | 00111101 | — | ) | |
| 42 | 00101010 | 052 | 2A | 00111111 | — | * | |
| 43 | 00101011 | 053 | 2B | 00111110 | — | + | |
| 44 | 00101100 | 054 | 2C | 00111010 | — | , | (Comma) |
| 45 | 00101101 | 055 | 2D | 00111011 | — | − | (Minus) |
| 46 | 00101110 | 056 | 2E | 00111001 | — | . | |
| 47 | 00101111 | 057 | 2F | 00111000 | — | / | |
| 48 | 00110000 | 060 | 30 | 00101000 | — | 0 | |
| 49 | 00110001 | 061 | 31 | 00101001 | — | 1 | |
| 50 | 00110010 | 062 | 32 | 00101011 | — | 2 | |
| 51 | 00110011 | 063 | 33 | 00101010 | — | 3 | |
| 52 | 00110100 | 064 | 34 | 00101110 | — | 4 | |
| 53 | 00110101 | 065 | 35 | 00101111 | — | 5 | |
| 54 | 00110110 | 066 | 36 | 00101101 | — | 6 | |
| 55 | 00110111 | 067 | 37 | 00101100 | — | 7 | |
| 56 | 00111000 | 070 | 38 | 00100100 | — | 8 | |
| 57 | 00111001 | 071 | 39 | 00100101 | — | 9 | |
| 58 | 00111010 | 072 | 3A | 00100111 | — | : | |
| 59 | 00111011 | 073 | 3B | 00100110 | — | ; | |
| 60 | 00111100 | 074 | 3C | 00100010 | — | < | |
| 61 | 00111101 | 075 | 3D | 00100011 | — | = | |
| 62 | 00111110 | 076 | 3E | 00100001 | — | > | |
| 63 | 00111111 | 077 | 3F | 00100000 | — | ? | |
| 64 | 01000000 | 100 | 40 | 01100000 | — | @ | |
| 65 | 01000001 | 101 | 41 | 01100001 | — | A | |
| 66 | 01000010 | 102 | 42 | 01100011 | — | B | |
| 67 | 01000011 | 103 | 43 | 01100010 | — | C | |
| 68 | 01000100 | 104 | 44 | 01100110 | — | D | |
| 69 | 01000101 | 105 | 45 | 01100111 | — | E | |
| 70 | 01000110 | 106 | 46 | 01100101 | — | F | |
| 71 | 01000111 | 107 | 47 | 01100100 | — | G | |
| 72 | 01001000 | 110 | 48 | 01101100 | — | H | |
| 73 | 01001001 | 111 | 49 | 01101101 | — | I | |
| 74 | 01001010 | 112 | 4A | 01101111 | — | J | |
| 75 | 01001011 | 113 | 4B | 01101110 | — | K | |
| 76 | 01001100 | 114 | 4C | 01101010 | — | L | |
| 77 | 01001101 | 115 | 4D | 01101011 | — | M | |

## Appendix 1 (*cont.*)

| Decimal | Binary | Octal | Hex | Gray | BCD | ASCII | |
|---------|--------|-------|-----|------|-----|-------|---|
| 78 | 01001110 | 116 | 4E | 01101001 | — | N | |
| 79 | 01001111 | 117 | 4F | 01101000 | — | O | |
| 80 | 01010000 | 120 | 50 | 01111000 | — | P | |
| 81 | 01010001 | 121 | 51 | 01111001 | — | Q | |
| 82 | 01010010 | 122 | 52 | 01111011 | — | R | |
| 83 | 01010011 | 123 | 53 | 01111010 | — | S | |
| 84 | 01010100 | 124 | 54 | 01111110 | — | T | |
| 85 | 01010101 | 125 | 55 | 01111111 | — | U | |
| 86 | 01010110 | 126 | 56 | 01111101 | — | V | |
| 87 | 01010111 | 127 | 57 | 01111100 | — | W | |
| 88 | 01011000 | 130 | 58 | 01110100 | — | X | |
| 89 | 01011001 | 131 | 59 | 01110101 | — | Y | |
| 90 | 01011010 | 132 | 5A | 01110111 | — | Z | |
| 91 | 01011011 | 133 | 5B | 01110110 | — | [ | |
| 92 | 01011100 | 134 | 5C | 01110010 | — | \ | |
| 93 | 01011101 | 135 | 5D | 01110011 | — | ] | |
| 94 | 01011110 | 136 | 5E | 01110001 | — | ^ | |
| 95 | 01011111 | 137 | 5F | 01110000 | — | __ | (Underline) |
| 96 | 01100000 | 140 | 60 | 01010000 | — | ` | (Accent Grave) |
| 97 | 01100001 | 141 | 61 | 01010001 | — | a | |
| 98 | 01100010 | 142 | 62 | 01010011 | — | b | |
| 99 | 01100011 | 143 | 63 | 01010010 | — | c | |
| 100 | 01100100 | 144 | 64 | 01010110 | — | d | |
| 101 | 01100101 | 145 | 65 | 01010111 | — | e | |
| 102 | 01100110 | 146 | 66 | 01010101 | — | f | |
| 103 | 01100111 | 147 | 67 | 01010100 | — | g | |
| 104 | 01101000 | 150 | 68 | 01011100 | — | h | |
| 105 | 01101001 | 151 | 69 | 01011101 | — | i | |
| 106 | 01101010 | 152 | 6A | 01011111 | — | j | |
| 107 | 01101011 | 153 | 6B | 01011110 | — | k | |
| 108 | 01101100 | 154 | 6C | 01011010 | — | l | |
| 109 | 01101101 | 155 | 6D | 01011011 | — | m | |
| 110 | 01101110 | 156 | 6E | 01011001 | — | n | |
| 111 | 01101111 | 157 | 6F | 01011000 | — | o | |
| 112 | 01110000 | 160 | 70 | 01001000 | — | p | |
| 113 | 01110001 | 161 | 71 | 01001001 | — | q | |
| 114 | 01110010 | 162 | 72 | 01001011 | — | r | |
| 115 | 01110011 | 163 | 73 | 01001010 | — | s | |
| 116 | 01110100 | 164 | 74 | 01001110 | — | t | |
| 117 | 01110101 | 165 | 75 | 01001111 | — | u | |

## Appendix 1 (*cont.*)

| Decimal | Binary | Octal | Hex | Gray | BCD | ASCII | |
|---------|--------|-------|-----|------|-----|-------|---|
| 118 | 01110110 | 166 | 76 | 01001101 | — | v | |
| 119 | 01110111 | 167 | 77 | 01001100 | — | w | |
| 120 | 01111000 | 170 | 78 | 01000100 | — | x | |
| 121 | 01111001 | 171 | 79 | 01000101 | — | y | |
| 122 | 01111010 | 172 | 7A | 01000111 | — | z | |
| 123 | 01111011 | 173 | 7B | 01000110 | — | { | |
| 124 | 01111100 | 174 | 7C | 01000010 | — | \| | |
| 125 | 01111101 | 175 | 7D | 01000011 | — | } | |
| 126 | 01111110 | 176 | 7E | 01000001 | — | ~ | |
| 127 | 01111111 | 177 | 7F | 01000000 | — | DEL | (Delete, rubout) |
| 128 | 10000000 | 200 | 80 | 11000000 | — | — | (Undefined in ASCII) |

## Appendix 2: Metric Prefixes

| Prefix | Abbreviation | Multiplier |
|--------|-------------|------------|
| atto- | a- | $10^{-18}$ |
| femto- | f- | $10^{-15}$ |
| pico- | p- | $10^{-12}$ |
| nano- | n- | $10^{-9}$ |
| micro- | u- | $10^{-6}$ |
| milli- | m- | $10^{-3}$ |
| centi- | c- | $10^{-2}$ |
| deci- | d- | $10^{-1}$ |
| deka- | da- | $10^{1}$ |
| hecto- | h- | $10^{2}$ |
| kilo- | k- | $10^{3}$ |
| mega- | M- | $10^{6}$ |
| giga- | G- | $10^{9}$ |
| tera- | T- | $10^{12}$ |
| peta- | P- | $10^{15}$ |
| exa- | E- | $10^{18}$ |

# Index/Glossary

Absolute Address   The actual, physical address of a memory location.
See also Indexed Addressing, Indirect Addressing

Access Time   Time required to read a piece of data from memory.
vs. Instruction Time, 33

Accumulator   A register loaded with the results of an operation

A/D Converter   (Analog-to-Digital Converter), 18

Address Space   The number of locations that can be addressed, usually a function of the number of bits devoted to the address; 33

Addressing, 14

Algorithm   An unambiguous, clear procedure; 28, 30, 39

Alpha Particle   The nucleus of a helium atom (two protons plus neutrons). An alpha particle can damage data stored in a dynamic RAM; 44

ALU (Arithmetic Logic Unit)   The section of a processor that performs arithmetic and logic functions; 10

Amdahl, Gene M., 71

Analog, 7
computer, 21
devices, 18
vs. digital, 16, 17, 21, 41
See also Digital Hardware Design

ANSI (American National Standards Institute)   A standards-setting organization; 9
See also Y32.14

AQL (Acceptance Quality Level)   The maximum permitted percentage of faulty parts; 71

Architecture   The high-level design of a project; 28, 31, 38

Arithmetic, 39–41
See also ALU, Carry-Save Adder

Arrows (in drawings), 111

ASCII (American Standard Code for Information Interchange)   A widely used encoding of alphanumeric characters into numbers; 9, Appendix 1

Assembly Language, 15, 23

Asynchronous   (1) A design method that is not controlled by a master clock; 13, 54 (2) Two processes are asynchronous with respect to each other if they do not share a common clock.
See also Combinatorial

ATP (Acceptance Test Procedure), 103–4

Availability   A factor between 0 and 1 that indicates a system's worthiness (the closer to 1.0, the better). Availability is computed MTBF/(MTBF + MTTR).
See also MTBF, MTTR

Average Case vs. Worst Case, 39

Background Task   See Foreground Task.

Backplane   The part of a system's chassis that carries the signals to interconnect the modules of the system and provides the modules' power and ground connections; 24, 55, 89

**127**

Bandwidth   The capacity for performing a periodic task. For example, if an adder is being operated at 40 MHz, its bandwidth is 40 MHz. If every 5th add is used for process A, we can say that process A takes 20% of the adder's available bandwidth.
See also Frequency

Base (number), 7

Baseboard   The hard surface (usually fiberglass) that supports devices (such as chips and capacitors) and their interconnecting wires; 24
See also Circuit Board

Bathtub Curve   A failure-rate curve showing failures vs. time that indicates higher failures during the early and late parts of a device's life, 43

Baud   Serial transmission rate, measured in bits per second.

BCD (Binary Coded Decimal)   An encoding of base 10 numbers with 4 bits used for each digit; 9, Appendix 1
See also Counter (decade)

BCH Code, 45

Benchmark   A standard test for measuring performance (usually speed).
See also MIPS, MFLOPS

Bends (in wires), 97

Bergeron Diagram   Also called a line reflection diagram, this technique graphically describes the voltage of a signal being reflected due to impedance mismatch.
See also Reflections, Ringing

Beta Site   A customer willing to evaluate a new product which is expected to still contain a few bugs.

Binary   Base 2; 7

Bipolar   Containing transistors such as those in the TTL and ECL logic families.
See also ECL, TTL, Unipolar

Bit, 7

BIT (Built-In Test)   Facilities added to a design to enable some error checking.
See also Testability

Bit-Slice Processor   A processor designed to be easily expanded in width. The AMD 2901 is a popular 4-bit-wide example; 20, 23

Bit-Vector Machines   A machine with a very wide data word (hundreds to thousands of bits wide) and provided with only logic operations. The extreme parallelism produces considerable speed; 41

Black Box   An entity whose function is known but whose algorithm is unknown or not important; 9, 13, 18, 30, 52

Block Diagram, 28, 111

Block Floating Point   An implementation of real numbers similar to floating point except that a single exponent is common to a group (block) of many mantissas. Block floating point is less general than floating point because each number shares a common exponent, but it uses less memory and can provide faster arithmetic.

Boole, George, 4

Boolean Algebra   The algebra of 2-valued (binary) logic; 4–6, 55

Bootstrap   Typically, a tiny computer program capable of little more than loading into memory a larger, more capable program.

Boundary Error   An error occurring when an extreme input is used; 37

Bubble Memory   A type of serial memory, characterized by being nonvolatile and relatively slow; 19, 38

Buffer   A one-input gate which outputs what is input (like a noninverting inverter). It is used to increase fanout or otherwise change the characteristics of the signal.

Burn in   To operate a device, often in a harsh environment, with the hope of detecting early failures; 43–4

Bus   A collection of signal lines which all operate together to transmit a group of related signals. Often the

Compute-Bound   Limited by the speed of processing, instead of being limited by I/O speed; 33
See also I/O Bound
Computer, 14–16, 20, 23
See also Bit-Slice Processor, Microcomputer, Microprogramming
Concatenate, 18
Conceptual Integrity, 32
Configurations, Choice of, 38–9
Connectors, 96
Constants, Allocation of, 34
Continued Fraction   A representation of a real number which has some computational advantages over the more common positional notation; 40
Continuity Tester, 24
Contradictory Advice, xi, 27
Control Lines   The controlling inputs to a device such as clocks and commands (other inputs might be data or address lines); 54
Control Store   Memory used to hold microcode; 38
See also Harvard Architecture, Microprogramming
Cooling, 91
See also Heat, Fans
CORDIC (COordinate Rotation DIgital Computer) Algorithm   An elegant and simple algorithm for performing many arithmetic, trigonometric, and logarithmic functions with only adds and table lookups; 39
Core   (1) A nonvolatile memory which stores bits as magnetic fields in tiny rings; 19, 38 (2) A casual term for a computer's primary solid-state (non-rotating) memory.
Costs, 55, 75, 89
Counter, 13, 17, 35, 114
Counter, Decade   A mod-10 counter. The outputs are BCD encodings of base 10 values; 13
Counter, Gray   A counter that sequences through the Gray code; 35–6
See also Gray Code
Counter, Johnson   A mod-2n counter which cycles through the codes

shown in Table G–1. An n-bit Johnson counter has 2n states. A shift register is often used to implement this sequence (also called a moebius counter or switchtail counter); 67

| Index | Johnson | Ring |
|-------|---------|------|
| 0     | 0000    | 0001 |
| 1     | 0001    | 0010 |
| 2     | 0011    | 0100 |
| 3     | 0111    | 1000 |
| 4     | 1111    | 0001 |
| 5     | 1110    | 0010 |
| 6     | 1100    | 0100 |
| 7     | 1000    | 1000 |
| 8     | 0000    | 0001 |

**Table G–1:** Successive values from 4-bit-wide Johnson and ring counter sequences.

Counter, Modulus   A counter with a certain number of states. A mod-n counter has n states. For example, a mod-3 counter will count: 00, 01, 10, 00, 01, . . . . Binary counters are mod-n counters with n always a power of 2. A mod-n counter is also called a divide-by-n counter.
Counter, Ring   A counter which cycles through the codes shown in Table G–1. An n-bit ring counter has n states. Like a Johnson counter, a ring counter is often implemented with a shift register (also called a ringtail shift register).
Coupling   (1) A module is closely coupled with another if to understand one, you must understand the other. Usually, coupling between modules should be minimized; 30 (2) Modules are closely coupled if they are tightly synchronized. Loosely coupled modules interact but operate autonomously.

CPU (Central Processing Unit) The part of a computer that executes the software.

CRC (Cyclic Redundancy Check) A method of tagging a serial data stream so that most transmission errors are noticed; 45, 46

Crosstalk A disturbance in the signal on one wire by the signal on another wire; 17, 88, 97

Crystal, 18, 102–3

Custom Logic, 20, 52–3

Cycle Stealing Overriding the operation of the processor for a short time. For example, a data storage circuit may cycle steal from the processor so that it can service a real-time DMA input. A cycle steal is similar to an interrupt except that the state of the machine need not be stored.

Cycle Time The minimum time between two memory accesses, at least as long as the access time. For example, core memory and dynamic RAMs have cycle times considerably longer than their access times.

DAC (Digital-to-Analog Converter), 18

Daisy-Chain To connect from one point to the next, to the next, and so forth; 34, 77, 91

Data Book, 11, 25, 48, 111

Dataflow Architecture A parallel, non von Neumann architecture in which data is operated on as it passes from processor to processor.

Date Code A code put on a device by the manufacturer identifying the batch from which the device came; 106

Datum A single piece of data (pl. data).

De Morgan's Theorem, 10–11

Debugging, 99–108
  and communication, 37
  and modularity, 32
  and packaging, 93
  and signature analysis, 47
  and silkscreen, 94

state generators, 35
and static, 69
tools, 23–4, 101
tristate busses, 55

Decoupling Capacitor A capacitor used to dampen noise in the power and ground planes; 38, 44, 69, 81–2

Delay Line A device which delays a signal; 56

Demultiplexer, 10, 17

Derate To reduce the actual stresses on a device below the worst-case stress allowed by the manuacturer; 43, 71, 72–3

Design
  high-level, 27–9
  languages, 110
  reviews, 101–2
  See also Digital Hardware (design)

Diagnostics, 42

Die An integrated circuit before it is placed in its package (pl. dice).

Differential Signals A pair of true and complement signals which transmit a signal not by their absolute voltages but by the *difference* between the voltages; 65, 66, 79

Digit On-Line Arithmetic, 40

Digital Hardware
  design, vii, 16–20
  vs. analog, 21, 41
  vs. logic, 2, 20
  parallelism of, 22
  vs. software, xi

DIL (Dual In-Line) See DIP.

DIP (Dual In-line Package) An integrated circuit packaging technique in which the input/output signals are conducted by two rows of leads which are perpendicular to the package; 92
  clips, 24

Distributor A company which sells products to users, similar to a retailer in the food industry; a middleman; 71

Divide, 41

DMA (Direct Memory Access) The accessing of memory by a peripheral

DMA (Direct Memory Access) (*cont.*) device without the intervention of the processor.

Documentation , 28, 30, 48, 50–1, 110–122
and compatibility, 29
of corrections, 100
and debugging notebook, 104
of PROMs and programmable logic, 52, 95
of signals, 70
and signature analysis, 47
for user, 34

Don't Care   A logic state which can be either high or low—neither state makes any difference to the function being performed; 5–7, 29

DRAM (Dynamic Random Access Memory)   A popular solid-state memory which is relatively inexpensive, dense, fairly fast, and volatile. The address is typically sent in two halves, called the row and column addresses; 18–19
array, 38
debugging, 102
and decoupling, 81
hierarchy, 37
initialization, 70
and memory scrubbing, 44
upgrades, 34

Driver   A buffer with unusually high capability for driving long signal lines; 82–3
See also Transceiver

DTL (Diode-Transistor Logic)   A predecessor to the TTL logic family rarely used today; 62, 63

Duplex, Full   A bidirectional communications path capable of being used in both directions concurrently.

Duplex, Half   A bidirectional communications path capable of being used in only one direction at a time.

Duty Cycle   The fraction of time during which a signal is active; 55

Dynamic RAM   See DRAM.

EBCDIC (Extended Binary Coded Decimal Interchange Code)   An 8-bit code, developed by IBM, used for encoding characters into numbers. Exists for the same purpose as ASCII but is one bit wider; 9

ECC (Error Correcting Codes)   A method of supplying redundancy in data so that errors can be detected and corrected; 23, 45

ECL (Emitter-Coupled Logic)   A bipolar logic family. Characterized by considerable speed and high power consumption; 65–6
and capacitance, 49
decoupling requirements, 81–2
and grounds, 70
and noise generation, 69
vs. other families, 61–3
and supply noise, 68

EDC (Error Detection and Correction)   See ECC.

Edge, Signal, 12, 54, 76

EEPROM (Electrically Erasable Programmable Read-Only Memory)   A ROM which can be electrically erased and reprogrammed; 19, 38

EMI/RFI (Electromagnetic Interference/ Radio-Frequency Interference)   Interference caused by high clock speeds and signal edge rates; 17
and errors, 44
and fiber optics, 80
limiting, 74
resonance, 77–8
and transceivers, 82

EMP (Electromagnetic Pulse)   An effect of a nuclear explosion which disrupts or destroys electrical devices; 80

Encryption, 37

Environment, 28

EPROM (Erasable Programmable Read Only Memory)   A PROM which can be erased and reprogrammed; 19, 38
See also EEPROM, UVPROM

Ergonomics   The science of making the machine fit the user, instead of the other way around; 33–4, 75

Foreground Task (*cont.*)
cuted when the foreground task is idle.

Form Factor   The size and shape of a circuit board; 89

Forth, 24

Forward Differences, 41

Frequency (content of signals), 76

Function, 2–3

GaAs (Gallium Arsenide)   A very fast new logic fabrication process. GaAs devices are much faster than ECL devices; 63

Gate, 2, 17, 55

giga-   Metric prefix meaning "x 10$^9$." The abbreviated prefix is "G."

Glitch   An unintentional narrow state change or spike on a signal; 35, 36, 44

Glue   Simple SSI and MSI logic used to get more complex LSI or VLSI components to work together.

Gray Code   A counting scheme which changes exactly one bit in the transition from one value to its successor; 5, 9, 35–6, 56, Appendix 1

Hamming Code   An EDC technique which corrects all 1-bit errors and detects all 2-bit errors; 45

Handshaking   A communications technique in which transfer requests and acknowledgements are sent to ensure correct transmission.

Hardwired   Fixed in hardware, as opposed to being in software.
See also Software

Harmonic   A signal whose frequency is an integer multiple of a given frequency. For example, digital clock signals contain  harmonics of the clock frequency; 56

Hartley   A unit of information content. One hartley = $\log_2 10$ bits = approximately 3.32 bits and is the number of bits required to store a single base 10 (BCD) value.

Harvard Architecture   An enhancement of the von Neumann computer architecture in which instructions and data are kept in separate memories. A modified Harvard architecture permits data transfers between the two memories; 33

Heat, 73, 93
See also Cooling, Temperature

Heat Sink   A structure for drawing away and dissipating heat; 72, 75, 95

Henry   A unit of inductance (metric symbol: H). Like a farad, a henry is very large and more commonly used units of inductance are microhenries and picohenries.

Hermetic   Sealed; airtight.

Hexadecimal   Base 16; 8, Appendix 1

HiNIL, 61–2

Hold Time   The amount of time an input to a synchronous device must remain stable after the clock edge arrives; 16, 50

Horizontal   A trait of microcode. The opposite of "vertical," horizontal microcode is wider for an identical application because it is unencoded. Vertical microcode uses 3 bits to control 1 of 8 different things, but horizontal microcode uses 8 bits to give the user complete simultaneous control of all 8 things. More parallelism is possible with the horizontal format, but microcode words are wider. In practice, an architecture usually uses some of each format.

HTL, 61–2

Hum   60 Hz noise, so called because in audio applications that frequency sounds like a hum.

Human Factors   See Ergonomics.

Hybrid   A single device composed of an interconnection of several integrated circuit dice.
See also Monolithic

Hysteresis   A noise-avoidance technique for digital signal inputs in which the low-to-high threshold transition is higher than the high-to-low threshold.

Hz (Hertz)  Cycles per second. Often used to mean "events per second."

IC (Integrated Circuit)  An electronic circuit fabricated on a semiconductor crystal. Often used to mean a packaged integrated circuit.

ICE (In-Circuit Emulation)  Mimicking a microprocessor with a tool which performs the same function as the microprocessor but which provides considerable insight into the internal state of the processor. This test tool is often used in debugging microprocessor-based systems; 75

IEEE (Institute of Electrical and Electronics Engineers)  An association of electrical and electronics engineers known for its publications; 9

IFL (Integrated Fuse Logic)  A type of programmable logic; 51, 52

IIL (Integrated Injection Logic)  A logic fabrication process characterized by low power dissipation and good packing density; 61–2
See also Compute-Bound

Impedance Mismatch, 17, 76–7, 97
See also Reflections, Ringing

Incremental Methods  Techniques whereby an expression can be quickly evaluated at a value by using the result of the evaluation at a nearby value; 41

Indexed Addressing  The content of an index register is added to an offset to arrive at an absolute address.

Indirect Addressing  The addressed location holds the *address* of the data, rather than the data itself.

Information Hiding, 30

Initialization, 38–9, 70

Insertion, Chip, 96

Instruction Set, 14, 15

Interleaved Memory  Memory which is divided into sections such that sequential memory requests will commonly be made to sequential sections. This speeds memory accesses; 38

Interrupt, 15, 19, 32

I/O (Input/Output).

I/O-Bound  Limited by I/O speed, instead of by the speed of processing; 33
See also Compute-Bound

IPL (Initial Program Load)  See Bootstrap

JAN (Joint Army-Navy).

JEDEC (Joint Electron Devices Engineering Council)  A standards-making body for integrated circuit packages.
See also Packaging

Josephson Junctions, 63

Jumper  (1) A wire which customizes a board to a particular application; 38 (2) a wire used to correct a circuit board flaw.

Junction  The interface which forms a transistor.

K  Abbreviation for 1024. For example, 2K bytes = 2048 bytes; 8

Karnaugh Map, 5–7, 55–6

KHz (kilohertz)  Thousands of cycles per second.

kilo-  Metric abbreviation for "x $10^3$." The abbreviated prefix is "k"; 8

Labels (signal), 115

Language
  hardware, 110
  software, 15, 23

Latch  A device with state which is loaded by a clock level, not a clock edge. The output follows (is the same as) the input when the clock is in one state and is frozen when the clock is in the other state; 11, 12, 54

Latch-up, 65

Layout, Board, 91–2

LCC (Leadless Chip Carrier)  A small IC packaging technique; 93

LCD (Liquid Crystal Display)  A display device requiring reflected light to be

LCD (Liquid Crystal Display) (*cont.*)
visible, often used in digital watches.
Usually miniaturized; 96

Lead  A pin on an integrated circuit
package; 92–3

LED (Light Emitting Diode)  A display
device which provides its own light.
Usually miniaturized; 96

Lettering, 112

Level-to-Pulse Converter, 57

Line Reflection Diagram  See Bergeron
Diagram.

Linear Device  An analog device.

Load  A destination for a signal.
See also Fanout

Logic, 2–4, 16, 20

Logic Analyzer, 24

Logic Probe, 24

Logic Simplification, 4–7, 55–6

Lookup Table  A RAM which holds the
values of a function. The memory
address (the argument to the function) looks up the corresponding
function value. Functions such as
square root, sine, or log can be computed this way.

LSB (Least-Significant Bit)  Least significant data bit in a word; 7

LSI (Large-Scale Integration)  Devices
having between 100 and 1000 equivalent gates; 17, 53

Machine Language, 15, 23

Maintainability  A measure of the ease
with which the system is repaired;
49, 55, 111
See also MTTR

Man-Machine Interface (MMI)  The fa-
cade of the machine seen by the
user. Often used to describe the
software which interacts with the
user; 33

Mantissa  The fractional part of a floating point number (the other part is
the exponent).

Manufacturer, 25
errors from, 71
higher-quality parts, 51

specifications from, 16, 49, 50, 71, 72
standard products, 17, 20
See also Second Sources, Data Book

Marketability, 28

Mask, PC, 87

Medium  Nonvolatile device used to
store digital data; for example, magnetic disk or tape, optical disc,
punched cards, or paper tape (pl.
media).

mega-  Metric prefix meaning "x $10^6$."
The abbreviated prefix is "M."

Memory, 13–4, 17, 18–20
hierarchy, 23, 37
market for, 62–3, 64
scrubbing, 44, 45
size, 33

Memory-Mapped I/O  Memory addresses assigned to peripheral I/O
ports. This architecture simplifies
interfacing with peripherals.

Metastability, 43, 44, 51, 106

Metric Prefixes, Appendix 2

MFLOPS (Millions of Floating-point Operations Per Second)  A unit of measurement used in comparing the
speeds of computers, pronounced
"megaflops."
See also MIPS

MHz (megahertz)  Millions of cycles per
second.
See also Hz

micro-  Metric prefix meaning "x $10^{-6}$."
The abbreviated prefix is "$\mu$."

Microcode  Software which controls a
processor. On a computer, microcode is one level lower (closer to the
hardware) than machine language;
23

Microcomputer  A computer built from
a small number of devices; 20, 31

Microprocessor  The processing portion
of a microcomputer; 17, 20, 55
and clocks, 102
market for, 62–3
and MOS, 64

Microprogramming, 23

Mil  1/1000 inch.

MIL-HDBK-217C  Military device failure prediction guideline; 71

milli-  Metric prefix meaning "x $10^{-3}$." The abbreviated prefix is "m."

MIL–STD–806B  Military logic drafting standard (superseded by ANSI Y32.14); 113

MIL–STD–883B  Military testing standard for high-reliability parts.

MIMD, 16

Minimization, 4–7, 55–6

MIPS (Millions of Instructions Per Second).  Called also millions of operations per second (MOPS), this is a unit of measurement used in comparing the speeds of computers.
See also MFLOPS

Mixed Logic, 10

MNOS (Metal Nitride Oxide Silicon)  A type of logic used in making nonvolatile memory.

Modular Arithmetic  A type of arithmetic which uses parallelism to boost speed; 40

Module  A portion of a system; 30, 32, 90, 110

Monolithic  Fabricated with a single integrated circuit. Hybrid circuits use more than one integrated circuit.
See also Hybrid

Monostable, 18

MOPS  See MIPS

MOS (Metal Oxide Semiconductor)  A type of logic characterized by low power consumption and high packing density; 38, 53, 64–5
See also CMOS

MSB (Most-significant bit)  Most significant data bit in a word; 7

MSI (Medium-Scale Integration)  The density of devices having between 12 and 100 equivalent gates; 17

MTBF (Mean Time Between Failure)  The larger the MTBF, the more reliable the system; 45, 71, 100
See also Availability, Reliability

MTTR (Mean Time To Repair)  The smaller the MTTR, the faster the system can be repaired.
See also Availability, Maintainability

Multilayer, 24, 80–1, 87

Multiplexer, 10, 17, 21

Multiplexing  Alternating between two or more sources.

Multiwire, 24, 80–1, 87

nano-  Metric prefix meaning "x $10^{-9}$." Abbreviated prefix is "n."

NC (No Connect), 114, 117

Negative Logic, 10

Negative Numbers, 8–9

Nibble  Half a byte (4 bits). Also spelled "nybble."

NMOS (Negative Metal Oxide Semiconductor)  See MOS.

Noise, 41, 76–82
  on busses, 82
  and debugging, 99
  and hysteresis, 18
  in supplies, 38, 44

Noise Margin  The difference between an output logic level and the range for an acceptable input logic level; 55
See also Hysteresis

Nonvolatile Memory  A nonvolatile memory does not need power to retain its stored data; 14, 19, 38

Normalize  To left shift the mantissa (fraction) of a floating point number until the leftmost bit is a one. The exponent is correspondingly adjusted. Some floating point operations require normalized input.

Nuclear Radiation, 44, 88
See also EMP

Number System, 7–9

Occam's Razor  The belief that if two theories explain the same phenomenon, the simpler is more likely to be correct (attributed to William of Occam, ca. 1300–1349).

Octal  Base 8; 8, Appendix 1

PROM, 19, 38, 52, 55
  See also EEPROM, EPROM, UVPROM
Propagation Delay  The delay through a
    device from the triggering input sig-
    nal to the output signal; 16, 25, 49–
    50
  minimum, 48
  vs. temperature, 68
  worst-case, 50
Protocol  The convention established
    for communication.
Pullup  A resistor which encourages the
    signal on a line to remain high. A
    *pulldown* resistor tries to keep the
    signal low; 54
  on bus, 74
  drawing, 116
  TTL, 69, 73, 107
Pulse, 11, 49
Pulse-to-Level Converter, 57


Quality  The measure of the fraction of
    working parts shipped by a manu-
    facturer; 71
  See also Reliability
Queue, 19–20


Race Condition, 48
RAM (Random-Access Memory)  A
    memory in which every location can
    be accessed in equal time. This can
    be contrasted against serial memory,
    in which the successor to the most
    recently accessed location can be ac-
    cessed most quickly; 14, 18–9, 20
  See also DRAM, PROM, ROM, SRAM
Rational Arithmetic  A number repre-
    sentation scheme which postpones
    divisions and, hence, increases accu-
    racy; 41
Read-Mostly Memory  A nonvolatile
    memory which will tolerate a very
    high number of reads but a lesser
    number of writes; 38
Real Estate  Area on a circuit board.
    Designers often look for smaller

chips or more compact circuits
    which occupy less real estate.
  See also Footprint
Real-Time, 32
Receiver  A device designed to input
    data from a transmission line; 82–3
  See also Transceiver
Redundancy
  data, 45
  module, 46
Reflections, 17, 76, 105
  in bent wire, 97
  of clocks, 68
  and termination, 78
  See also Ringing
Refresh, DRAM, 18
Register, 13, 17, 83
  drawing, 114
  vs. latch, 54
  in pipeline, 21–2
Register File  A small, high-speed
    RAM perhaps 16–256 locations in
    size.
Regression Testing  Retesting a proto-
    type design using tests which have
    already been passed; 47, 99
Relative Address  An address which
    represents an offset from a base ad-
    dress.
Reliability  The failure rate of a device
    or system over time; 41, 71, 75
  See also MTBF, Quality
Repair, 48, 90
  See also MTTR
Reset, 39, 42
Resonance, 77
Ribbon Cable, 79
Ringing, 49, 76, 105
  See also Impedance Mismatch, Reflec-
    tion
RISC (Reduced Instruction Set Com-
    puter)  A newer philosophy in in-
    struction set design which
    encourages a few simple, fast in-
    structions instead of many complex,
    cumbersome instructions; 31
  See also CISC
Rise/Fall Time, 76

Tri-State  A logic output option which provides a turned-off state in addition to high and low. This allows several output devices to share a line. Also called 3-state; 21, 55, 67

Truncation  Dropping the least-significant bits of a number.

Truth Table, 2–3, 4, 5

TTL (Transistor-Transistor Logic)  A bipolar logic family characterized by a wide number of devices and fairly high speed; 61–5, 66

capacitance, 49

decoupling, 81–2

floating inputs, 107

mixing, 70

and noise, 68

power consumption, 38, 53

pullups, 69

Turing Machine  A hypothetical computer with an extremely simple instruction set; 14, 31

Twisted Pair  A pair of twisted wires, one of which carries a signal and the other a ground or complement signal; 79

See also Differential Signals

Two's Complement Notation  Binary notation for numbers in which negative values are one greater than the one's complement notation; 8

See also One's Complement Notation

UART (Universal Asynchronous Receiver/Transmitter)  A device which performs the asynchronous translation between serial and parallel data. Often used with communications streams.

ULSI (Ultra-Large-Scale Integration)  A possible term for device densities higher than VSLI.

See also VLSI

Underflow  The situation when an arithmetic instruction (usually divide) produces a result smaller than the smallest representable number.

Unipolar  Containing transistors such as those in the CMOS logic family.

See also Bipolar

Unusual Techniques, 39–40

UVPROM (Ultra Violet PROM)  A read-only memory which can be erased with exposure to ultraviolet light and reprogrammed; 19, 38

Variable (logic), 7

Vector  A large number of quantities which can be operated on in parallel.

See also Bit-Vector Machines

Vertical  See Horizontal.

VHSIC (Very High Speed Integrated Circuit)  A U.S. Department of Defense program for producing dense and fast new VLSI devices.

Via  A copper-plated hole which interconnects signals on two sides of a circuit board; 87

Video

and algorithms, 39

and ergonomics, 33

testing, 47, 101, 102, 103, 104

Virtual  Pretend, not actual. A system with virtual memory, for example, gives the user the impression of more solid-state memory than actually exists (in fact, slower storage holds the excess data).

VLSI (Very Large-Scale Integration)  The density of devices having over 1000 equivalent gates; 17, 53, 81, 90

See also ULSI

Volatile  The description of a memory which loses its data when the power is turned off; 14

von Neumann Architecture  The architecture used in most digital computers. This architecture has a single processor and control passes in a sequential fashion from instruction to instruction; 14, 15, 16